PSpice Text/Manual
An Introduction

James J. Tart

VANCE - GRANVILLE COMMUNITY COLLEGE

First Edition 1992
Revised Edition 1993
Second Edition 1994

Computer Programming And Teaching Publishers
C>PAT Publishers
Post Office Box 351
Henderson, NC 27536

Engineering Technology Series

Library of Congress Cataloging-In-Publication

Tart, James J.
PSpice Text/Manual-An Introduction/James J. Tart, A.A.S., B.S.

ISBN 0-9635788-1-2 1. Spice (Computer program) 2. Electric circuit analysis-Data processing. 1994

Computer>Programming And Teaching Publishers
P O Box 351
Henderson, NC 27536

To Elizabeth, Joette,
Julie and Kenneth...

...and to each student,

"Study to show thyself approved..."

THE SCIENTIFIC METHOD IN PSPICE EXPERIMENTATION

Instructor assigned and student created exercises may be used to reinforce concepts studied in each chapter. The latter being the most effective learning approach. Always use the scientific method where experimentation proves hypothesis. This method could be described as a step-by-step procedure using observation and experimentation in the solving of problems. In PSpice, the user moves from design expectations to experimentation to results. Steps in the scientific method are presented in the diagram below.

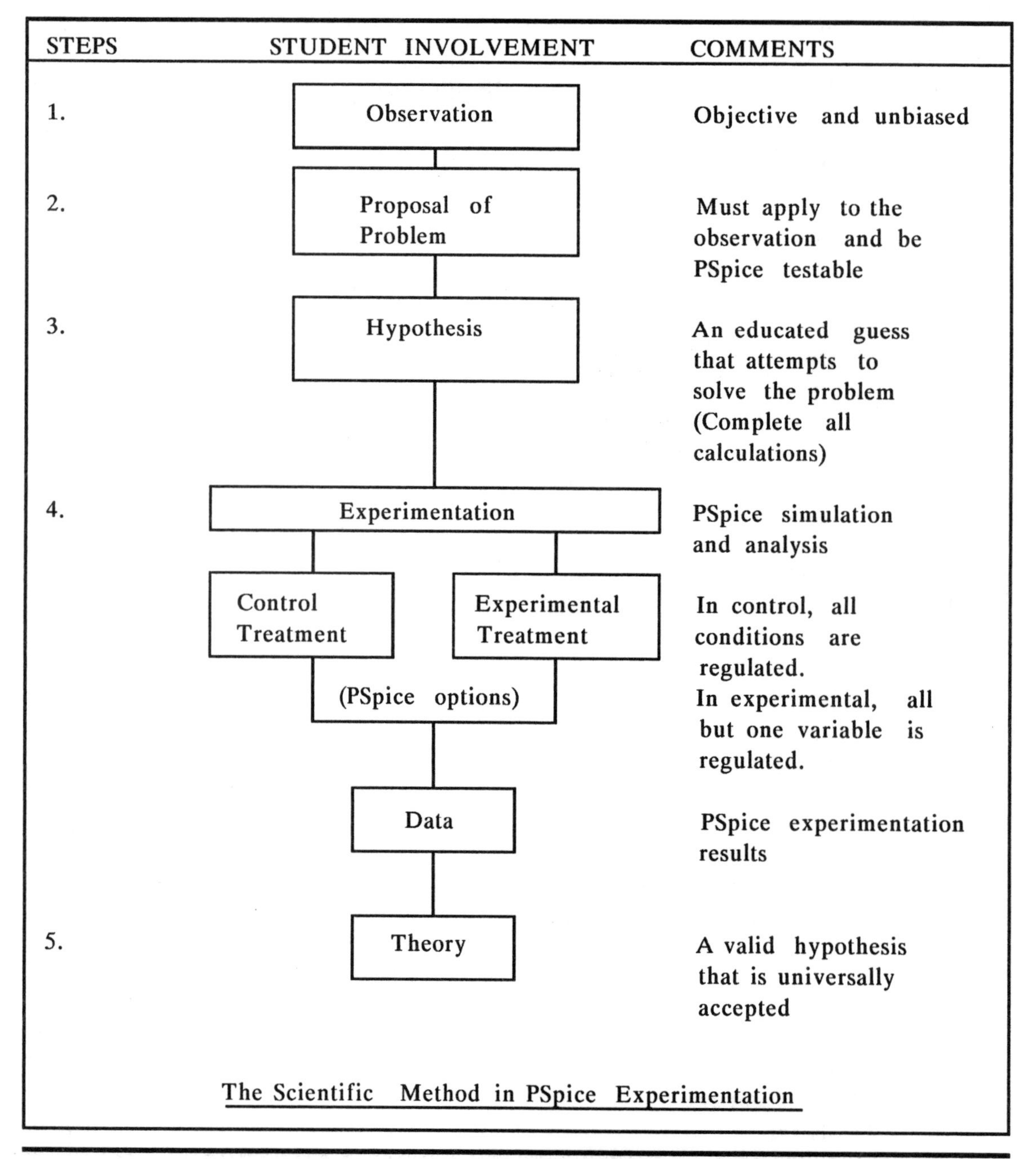
STEPS
STUDENT INVOLVEMENT
COMMENTS
1.
Observation
Objective and unbiased
2.
Proposal of Problem
Must apply to the observation and be PSpice testable
3.
Hypothesis
An educated guess that attempts to solve the problem (Complete all calculations)
4.
Experimentation
PSpice simulation and analysis
Control Treatment
Experimental Treatment
(PSpice options)
In control, all conditions are regulated.
In experimental, all but one variable is regulated.
Data
PSpice experimentation results
5.
Theory
A valid hypothesis that is universally accepted

The Scientific Method in PSpice Experimentation

CONTENTS

	Page
The Scientific Method in PSpice Experimentation	iv
Preface	vii
To the Student	x
Installing PSpice - Evaluation Version	xi
Chapter:	
1. Introduction to PSpice	1-1
2. Control Shell	2-1
3. Probe Graphical Waveform Analyzer "Oscilloscope" .DC Sweep	3-1
4. Stimulus Editor	4-1
5. Device Library	5-1
Parts Parameter Estimator	5-5
6. Series and Branch	
Series Circuits	6-1
Branch Currents	6-8
Independent Current Sources	6-10
7. Thevenin and Norton	
Thevenize with PSpice	7-1
Nortonize with PSpice	7-6
Component Models: RES CAP IND	7-9
Maximum Power Transfer; Impedance Matching	7-14
8. Rectifiers	
Full-Wave Rectifiers	8-1
Bridge Rectifiers	8-6
Using Transformers in PSpice	8-11
9. Transistor Curves	
BJT Collector Curves	9-1
JFET Drain and Transconductance Curves	9-6
Titles, Labels, and Arrows in Probe	9-4
10. Diode Circuits	
Voltage Doublers	10-1
Diode Limiters	10-3
DC Clampers	10-6
Peak-To-Peak Detectors	10-8
Zener Regulators	10-9

CONTENTS

Chapter:	Page
11. Bipolar Junction Transistor DC Bias and Class A Operation	11-1
12. Bipolar Junction Transistor AC Load Line Decibel Gain	12-1
13. Coupling and Bypass Circuits .AC Sweep - Small Signal Analysis	13-1
14. Resonance;	
Series Resonance	14-1
Parallel Resonance	14-6
RC - LR Phase Shift Circuits	14-9
Complex Numbers in AC Circuits	14-13
15. Series Regulator	15-1
16. Differential Amplifier Differential Gain, Common Mode Gain, CMRR	16-1
17. Linear Operational Amplifier	17-1
18. Linear Bandwidth Amplifier	18-1
19. Schmitt Trigger Circuits	
Inverting Schmitt Trigger	19-1
Non-inverting Schmitt Trigger	19-7
20. MORE PSPICE CIRCUITS	
Op Amp Relaxation Oscillator	20-1
JFET Amplifier	20-3
Colpitts Oscillator	20-5
Complementary Symmetry Amplifier	20-7
Active Low-Pass Filter	20-9
SETUPDEV.EXE Program	21-1
Complimentary Evaluation Software	21-2
Bibliography	22-1
Appendix A: SPICE Statements	23-1
Appendix B: Device Model Parameters And Subcircuit Statements	24-1

PREFACE

This PSpice Text/Manual was written in an effort to squeeze a PSpice introduction into our already cramped technology programs. The Engineering Technology dilemma of insufficient space for essential studies is constantly with us. Our students going into industry and those going on to four-year schools have special requirements for circuit simulation and analysis training. This challenge to upgrade our programs demands the best training available. PSpice, or a similar package, will open major possibilities for these students, and can not be ignored. For these and other reasons, we aimed for the minimum of a PSpice introduction that will provide that solid circuit simulation and analysis foundation needed by our students.

The overall objective of the Text/Manual is to be an accompaniment to fundamental electronic studies and intermediate circuit analysis. It may also be used as a one quarter course near the close of analog theory in engineering and vocational programs.

The Text/Manual is a tutorial/self-teach guide to circuit computer-aided design. PSpice lecturing should be minimum if experimentation follows concept studies in theoretical textbooks. Experimentation assignments can be as copious as is essential to reinforce previously studied theory.

An effort has been made to avoid crowding all PSpice program entry information into the beginning of the book. Additional PSpice uses and applications are introduced in succeeding chapters. Chapters build upon preceding chapters. Individual PSpice programs are introduced initially with specific circuit simulation and analysis following. In this format, the book can serve both as a source for PSpice introduction and as a learning tool.

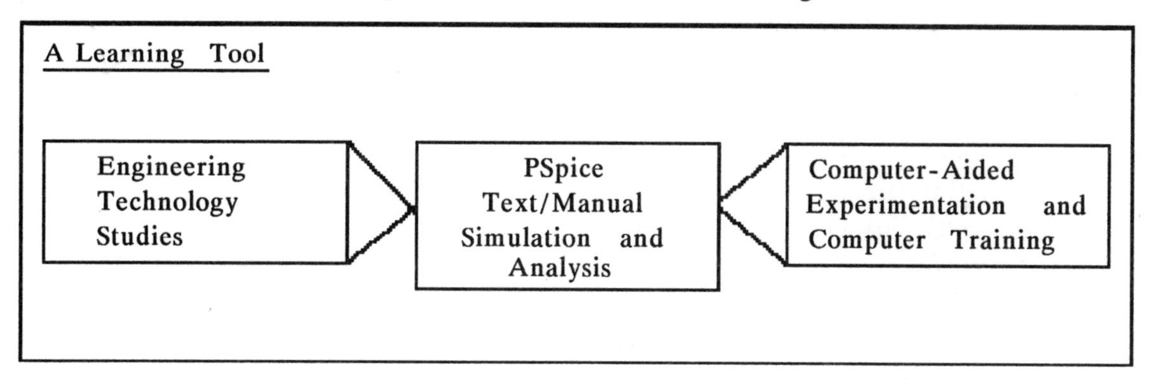

As students are introduced to new theory, computer-aided design follows in experimentation. The Text/Manual "learning tool" becomes a hands-on environment. For example, BJT and JFET transistor curves and load lines are studied in Chapter 9. In this chapter, the student is not only observing collector curves and load lines dynamically drawn but is also using the computer to enter the necessary equations as PSpice analyzes transistor operations.

The Text/Manual, although written for the Evaluation Version, is equally adaptable to a full production version of PSpice.

Evaluation Versions received from MicroSim Corporation are:

MicroSim Design System 2(C):

Evaluation - Version 4.03 program, 1/22/90,
(Reference 1988)

Evaluation - Version 4.04 program, 7/10/90

Evaluation - Version 4.04b program, 9/26/90

Evaluation - Version 4.05 program, 2/21/91

Evaluation - Version 5.0 program, 1991,

Evaluation - Version 5.2 program, 1992,

Evaluation - Version 5.3 program, 1993

Copying of the 5.2 Evaluation Version is welcomed and encouraged.

All PSpice software is the property of MicroSim Corporation, 20 Fairbanks, Irvine, CA 92718. Registered trademarks include:

PSpice(C)	Circuit Simulator
Probe(C)	Waveform Analyzer
StmEd(C)	Stimulus Generator
Parts(C)	Device Characterizer
PShell(C)	DOS Control Shell

The Install program for PSpice is licensed software of Knowledge Dynamics Corporation, P.O. Box 1558, Canyon Lake, Texas 78130-1558. Knowledge Dynamics Corporation owns by copyright and reserves all copyright protection worldwide; Copyright (C) 1987-1991. Install is provided to you for the exclusive purpose of installing PSpice. In no event will Knowledge Dynamic Corporation be able to provide any technical support for PSpice.

A PARTS Program Note

A possibility exist that the PARTS program in Chapter 5 may stall at the initial PARTS screen. It may be necessary to erase the files incorrectly created or modified in PARTS by the previous program user. For example, if the user "modified" files D1N4001.MDT and D1N4001.MOD while viewing the 1N4001 diode screens, and the modifications made the device unusable, then simply erase these two files at the PSEVAL52 directory and proceed.

In addition, all PARTS files should be stored in a separate directory after installing PSpice. These files can be copied back to the PSEVAL52 directory if needed. If hard drive space is available, the entire PSpice evaluation program can be installed to another directory called "PSSPARE". The instructor only should make these changes.

Special Recognition: University of California, Berkeley

Original SPICE development was centered in the Electrical Engineering and Computer Science Department, University of California, Berkeley. All United States SPICE programs of whatever name have the U.C. Berkeley foundational SPICE work as a base. U.C.'s publicly funded engineering efforts have served and will continue to serve the public and industry in electronic computer-aided design. Our special thanks for their outstanding contribution to SPICE-CAD systems.

PSpice, by MicroSim Corporation, was developed out of the U.C. Berkeley SPICE2 package, with many improvements. U.C.'c SPICE2G.6 circuit files (netlist) and simulation commands can run directly on the PSpice program.

PSpice Production Version Note

MicroSim offers discounts to educational institutions on most of their PSpice packages. If you are considering purchasing a full-scale version of PSpice, the following two notes are important. The Design Center - System 2 (menu driven), IBM-PC DOS/16M, production version of PSpice requires entended memory. The IBM-PC DOS(640k) low memory configuration does not support digital simulation and does not use extended memory.

Design Center - System 1 is Macintosh, SUN-4/Network, and HP9000/700/Network oriented. Design Center - System 3 is aimed at the IBM-PC/Windows and SUN-4/Network market.

Contact MicroSim for your requirement specifics.

Appreciation

I extend my appreciation to MicroSim Corporation for the use of the PSpice Evaluation Versions in the Text/Manual production that has grown to its present size of 280+ pages.

Also, a note of gratitude is due the administration of the college where I am privileged to teach for their patience while I wrote this edition. And very special thanks to my family for their continued support as I pursued a time consuming long-term goal.

Jim Tart

TO THE STUDENT

As technology expands, there are greater and greater possibilities of "gaps-in-learning" as education attempts to keep up with the knowledge explosion. Even now, Electronic Engineering Technology has several developing fields of specialization. Students in two-year restricted programs are not expected to study all specializations but some areas of expertise are considered indispensable for anyone entering engineering electronics. One of these areas of study is computer circuit simulation and analysis. PSpice Text/Manual offers foundational learning in this computer-aided design (CAD) field. In the future, you may build upon the PSpice introduction this book offers. You are given the opportunity to more than just "get your feet wet" in this important technology. You will receive significant CAD training in the PSpice Text/Manual that will insure against a rather significant "gap."

Circuit simulation and analysis is as new as the computer to our technology world. In the "old days," circuits were built from the breadboard up, from prototype through several stages of development and fine tuning. Designs were enhanced at the workbench with technicians grinding away hours of testing to prove specifications...changing components, the environment, stress tests, time tests, etc. After many weeks, perhaps months, an acceptable package would be processed with many dependability questions left unanswered. Today, many of these tests can be simulated with computer programs like PSpice, with much greater levels of accuracy. The time between concept to finished design and the level of excellence in design have been greatly improved. And on the way to this computer-aided design approach we have seen an impressive educational tool developed. This tool is available to you today with the PSpice Text/Manual. In this book you will find an introduction to PSpice's impressive circuit simulation and analysis program.

Happy Spiceing!

Installing PSpice - Evaluation Version

Software/Hardware Notes

PSpice analysis programs can be time consuming, even in the evaluation version. While waiting, it is a good idea not to press "Enter" a second time, or you may find yourself in an unfamiliar part of the program. PSpice remembers your "Enter(s)" and proceeds with the execution of your commands as if you remember also.

The speed of your computer is, of course, governed by the processor and coprocessor installed. If you are using a 286 without a coprocessor (minimum requirement), be prepared to wait a couple of minutes for analysis of some of the circuits in this Text/Manual. Time in the program is governed by the span and number of data points in a particular analysis.

MOUSE

The Arrow keys are quite convenient for moving around in PSpice programs and a mouse is not a must. But if you want speed and choose to use a mouse, Microsoft(C) compatibility is required.

Installation Information for:

MicroSim Design System 2(C) Evaluation - Version 5.2

Instructions for installing the Design System 2 Evaluation Version 5.2 are included on the disk. This section is provided to assist in determining system compatibility and as a guide to installation prompts.

Change to the root directory of your hard drive. For example, if drive C is your hard drive, type CD\ and Enter at the C:\ prompt.

Insert Diskette 1 of 2 into drive A.
Type "Install" at drive A DOS prompt, and press Enter.
The program will be copied to the boot drive (the drive with the AUTOEXEC.BAT and CONFIG.SYS files).

The system includes:

PSpice(C)	Circuit Simulator
Probe(C)	Waveform Analyzer
StmEd(C)	Stimulus Generator
Parts(C)	Device Characterizer
PShell(C)	DOS Control Shell

(Continued next page)

During the installation you will be asked to choose between one of two versions, depending on the configuration of your system:

System Version #1: Requires 640K of conventional memory and simulates analog circuits only. (3.8Mb of space required on hard drive. Additional space should be planned for your files)

System Version #2: Requires 1 Megabytes of extended memory for simulating mixed Analog/Digital circuits. (4.2Mb of space required on hard drive. Additional space should be planned for your files)

System Version #1 is suitable for all experimentation in the PSpice Text/Manual.

Select System Version #1 and follow directions. Select your boot drive. Accept PSEVAL52 as the directory and Enter.

Both the AUTOEXEC.BAT and CONFIG.SYS files may be modified to accommodate the Version you choose. The old files will be saved as backups.

C:\PSEVAL52 will be added (with your permission) to the path statement in the AUTOEXEC.BAT file. In addition, SET PSPICELIB=C:\PSEVAL52 will be placed in the AUTOEXEC.BAT file for efficient access to device libraries.

Choose "Go ahead and modify" unless you wish to make the modifications to the AUTOEXEC and CONFIG files later.

PROBE.EXE, PARTS.EXE, and STMED.EXE require a PSPICE.DEV file that specifies the display and print devices used in your system. The SETUPDEV program (p.20-1) can be executed later or entries can be made in the Control Shell to create the PSPICE.DEV file or you may choose to run it now. If Yes, Enter "Y".

Choose Display, Port, and Printer type. At Selection> Choose 4 and Enter to save if correct. At Selection> Choose 0 and Exit to leave SETUPDEV, the device file creation program. [If you change your system later (i.e.: monitor, printer), run SETUPDEV.EXE (p.20-1) to include changes]

The installation of MicroSim Design System 2 Evaluation - Version 5.2 is complete.

Reboot your computer to include changes in the AUTOEXEC.BAT and CONFIG.SYS files.

Note: Design System 2 Evaluation - Version 5.3 is now available and is completely compatible with the PSpice Text/Manual. For our purposes, the two versions are the same. Student Version - 5.4, with an EVAL.LIB library file, is also compatible. See page 21-2 for complimentary software.

Chapter 1

Introduction to PSpice

Objectives: (1) To introduce the circuit simulation idea
(2) To provide a PSpice Text/Manual overview
(3) To state the Text/Manual objectives
(4) To introduce a few PSpice program notes
(5) To furnish additional PSpice particulars

I. What Is Circuit Simulation And Analysis?

SPICE, "**S**imulation **P**rogram with **I**ntegrated **C**ircuit **E**mphasis," was first developed at the University of California, Berkeley. SPICE2 from this initial work is the foundation package for several commercial Spice programs. The PSpice package by MicroSim Corporation is an advanced version of this original circuit simulator and is the world leader in simulation and analysis programs. A note of appreciation is expressed to MicroSim for the use of their Evaluation program in putting together this Text/Manual.

Just a few years ago, students in electronics classes and many technicians in the work force spent most of their time with circuits and test equipment. Students spent their class time at work benches, rather than at desks and computers. Engineers and technicians on the job faced circuit development deadlines (and still do) with breadboards under their noses and wires strung out to power supplies, voltmeters, ammeters, waveform analyzers, signal generators, logic probes, and O'scopes. In this environment, developers and students were at home. A short time ago, all circuits were built before they could be tested. Indeed, the idea of production before extensive experimentation was unthinkable. But, electronics experimentation has been altered. Along with all the other tremendous changes that have taken place with the development of the computer, circuits no longer have to be constructed to be performance tested. In fact, more extensive testing and fine tuning can now be completed long before the production manager orders parts. The new approach is "computer circuit simulation", and the new testing is call "computer circuit analysis." Beginning students still spend their time at the workbench in the learning process, but in advanced training, and in the field, design has moved into the world of the microprocessor.

The computer program that has brought this change in the design and development environment in Spiceing. Using Spice, computers are programmed to:

1) Simulate the circuit,

2) Check out the initial feasibility of the circuit,

3) Test circuit operation in detailed analysis,

4) Run detailed testing variables on components and environment, and,

5) Provide analytical outputs in complex detail.

The time reduction in circuit modeling is staggering and the need for training in Spiceing is obvious. Our objectives in this Text/Manual are to introduce electronic students to this computer testing area of circuit development. So, put away your test leads for a while and explore this exciting new "workbench." But first, some fundamental thoughts about using PSpice or any other CAD package.

"PSpice does not totally take over design, but simulates previously designed circuits and analyzes circuit operation. In the first two years of electronics study, PSpice is used primarily to confirm textbook theory and problem solving calculations. As a learning tool, and as an engineering discipline, PSpice simulation of design and problem solving confirmations are particularly adaptable to The Scientific Method on page iv."

For these reasons, any circuit should be as complete as the designer or student can make it before running PSpice simulation and analysis. A well thought out design will include schematics, expected gains, power needs, impedances, component power dissipation, etc. PSpice will run a computerized model of the design from which fine tuning approaches may be made. Keep in mind that PSpice does not design circuits, but rather, is a tool for the designer and the student. A powerful tool!

To summarize, computer circuit simulation and analysis is continuing to explode. It is ever expanding. In other related areas, microelectronic chip fabrication is totally computerized and continues to grow in capacity and speed. Computer programs like AutoCad and OrCad are continuously expanding their roles in assisting drafting and fabrication layouts. In Spiceing, simulation software has taken the place of prototype and mid-stage workbench construction wherever possible. PSpice, the world leader in circuit simulation and analysis, has helped define this expanding field of computer-aided design. As a result of all this growth in technology, the requirements of electronics education has, of necessity, moved even faster and in a broader sense. This is our major area of interest.

This "knowledge explosion" at the technical level has resulted in the engineer's and technician's education and job becoming even more varied and complex. There are ongoing challenges in our colleges just to keep abreast of growth. This need for expansion is particularly acute at the community college engineering technology level. The addition of the PSpice Text/Manual to introduce computer assisted electronic circuit design is one of the efforts being made to meet these challenges.

Is PSpice difficult to learn?

What is PSpice CAD? How does it work? Is it difficult to learn? How much time is required to learn this computer program, PSpice? These and other questions, no doubt, come to mind. And the answers should be forthcoming with an overview of the Text/Manual. But first, let's take a look at a even bigger question about EET programs and PSpice and the future. If we were to attempt to tackle a detailed study of this extensive analytical tool in the first two years of study, we would need to dramatically alter our engineering technology programs. The problem is the lack of room in two-year programs. Programs are already full with entire courses dedicated to in-depth studies of engineering fundamentals, circuit analysis, and digital and microprocessor theory, with hardware circuit construction and/or analysis included in these courses. The point is...not one course can be eliminated..., and whatever is new must be in addition to core studies. As the electronics field expands into twenty-first century research and development, engineering technology students will need, at a minimum, computer-aided design fundamentals. Therefore, as a first objective, the Text/Manual is attempting a PSpice CAD introduction that will fit in with existing course materials.

> "...a first objective of the PSpice Text/Manual is...a PSpice CAD introduction that will fit in with existing courses..."

Now, to the question of difficulty in learning PSpice. The answer to "difficulty" is included in one of the above questions ... "How much time...?" Like all learning, "PSpice learning" is dramatically improved with time-on-task. And with PSpice, there are additional gains. The hours spent in the study of computer-aided design has at least two added benefits. One, the theory learned in several other courses will be positively reinforced. And two, PSpice trains on one of the most fascinating learning tools of our age...the computer. Instead of "how difficult?", begin with the thought that, for the electronics student,

"THE TIME SPENT WITH PSPICE WILL BE EXTREMELY BENEFICIAL AND SHOULD BE ENTERTAINING AND YES,...EVEN FUN."

A block diagram of PSpice operation from the Control Shell is presented on the next page.

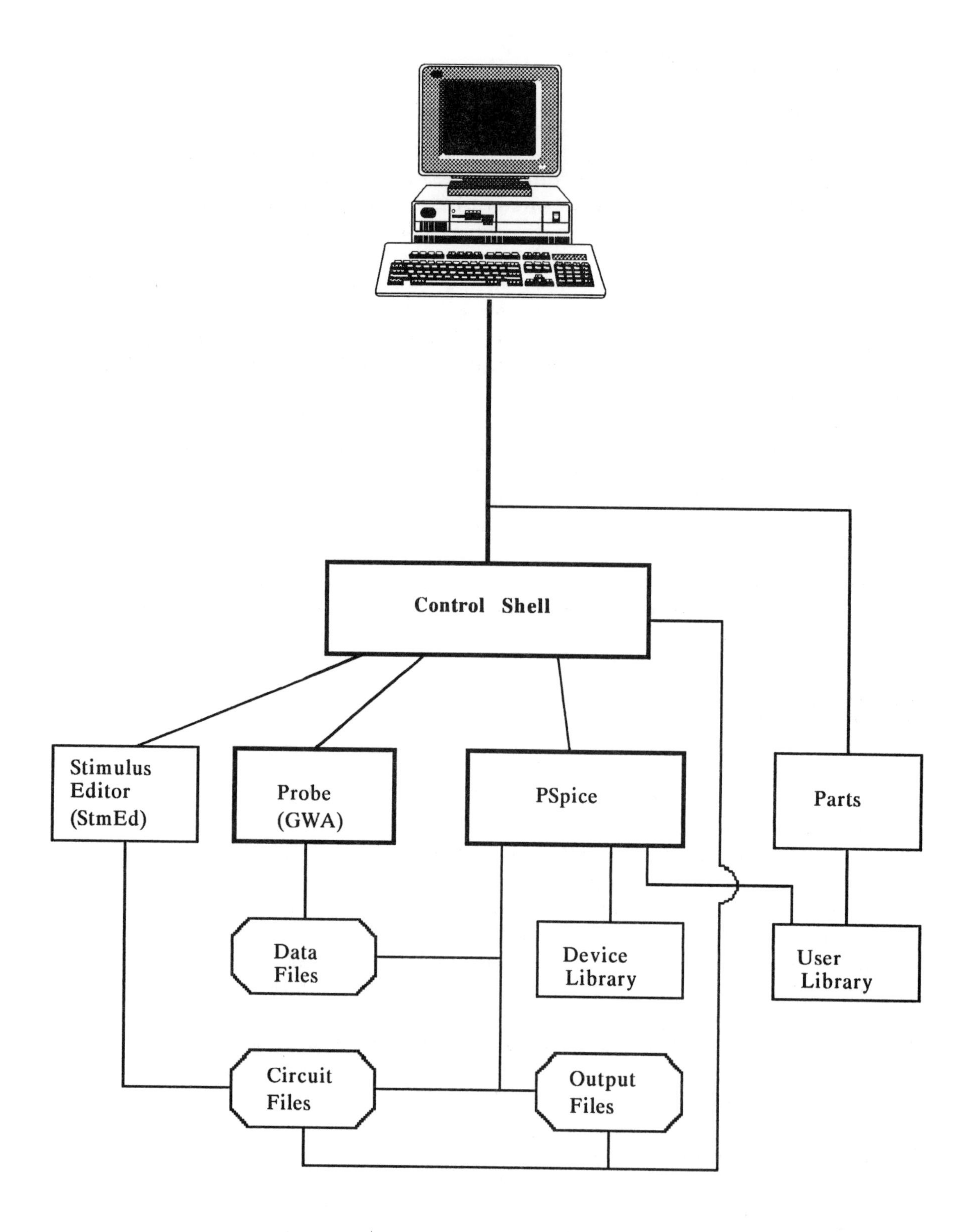

Figure 1. PSpice and The Control Shell

II. A PSpice Text/Manual Overview

Tutorial Approach

A second objective was achieved when the Text/Manual was written as a "self-teach, user-friendly guide." For example, the chapters are written in a step-by-step format before and after entering a circuit program. You, the student, learn by handling and repetition of handling the information before you. As a learner, you are placed into a hands-on environment. The central idea is to build self confidence in the operation of the PSpice program. The Text/Manual is written to accompany you in this building process. The experimentation included should follow regular textbook theoretical studies. Chapters can be repeated where difficulty is encountered. Confidence comes with accomplishment. Learning and growing are synonymous.

> "A second objective,...the Text/Manual was written as a "self-teach, user-friendly guide."

The Text/Manual builds chapter upon chapter. The Stimulus Editor, the Device Library, and the Parts program follow the introduction to the Control Shell and Probe. The remaining chapters are studies in using the PSpice program to simulate and analyze specific designs you have previously studied. You may create, or your instructor may assign, as many example circuits as necessary and analysis runs can be made to confirm theory.

Text/Manual Lesson Plans

The Control Shell

In Chapter 2, the PSpice Control Shell is introduced. This is the "home menu screen" for operating the program. (Figure 1) Options at the Control Shell pop-down menu screens include a Circuit Editor for writing and correcting the circuit description in PSpice language (the netlist). Device and component editing, input signal (stimulus) editing, simulation of the circuit, analysis of the circuit, viewing the Output file (.OUT), and Probe are also available through the Control Shell program.

Note: The PSpice program and other programs within PSpice can also be operated apart from the Control Shell by entering at the DOS prompt. See Figure 2.

Probe Graphical Waveform Analyzer

Chapter 3 presents Probe, the PSpice "Oscilloscope." More recently, Probe has been called the Graphical Waveform Analyzer. (Fig.1) PSpice data files (.DAT) are provided by the analysis program for use by Probe. It is at the Probe screens that the user sees the dramatic presentations of computer-assisted design. All the information that is needed for design, and more information than is needed, is meticulously presented by this fascinating piece of "test equipment." Probe is an independent program within PSpice and can be operated from the Control Shell or from DOS.

Stimulus Editor, The Input Generator

Chapter 4 is an introduction to the Stimulus Editor, or StmEd. The circuit stimulus, or input signal, can be created or adjusted to circuit specifications in StmEd. This "input generator" is also an independent program within PSpice and can be operated from the Control Shell or as a "free-standing generator" from DOS. See Figure 2 below.

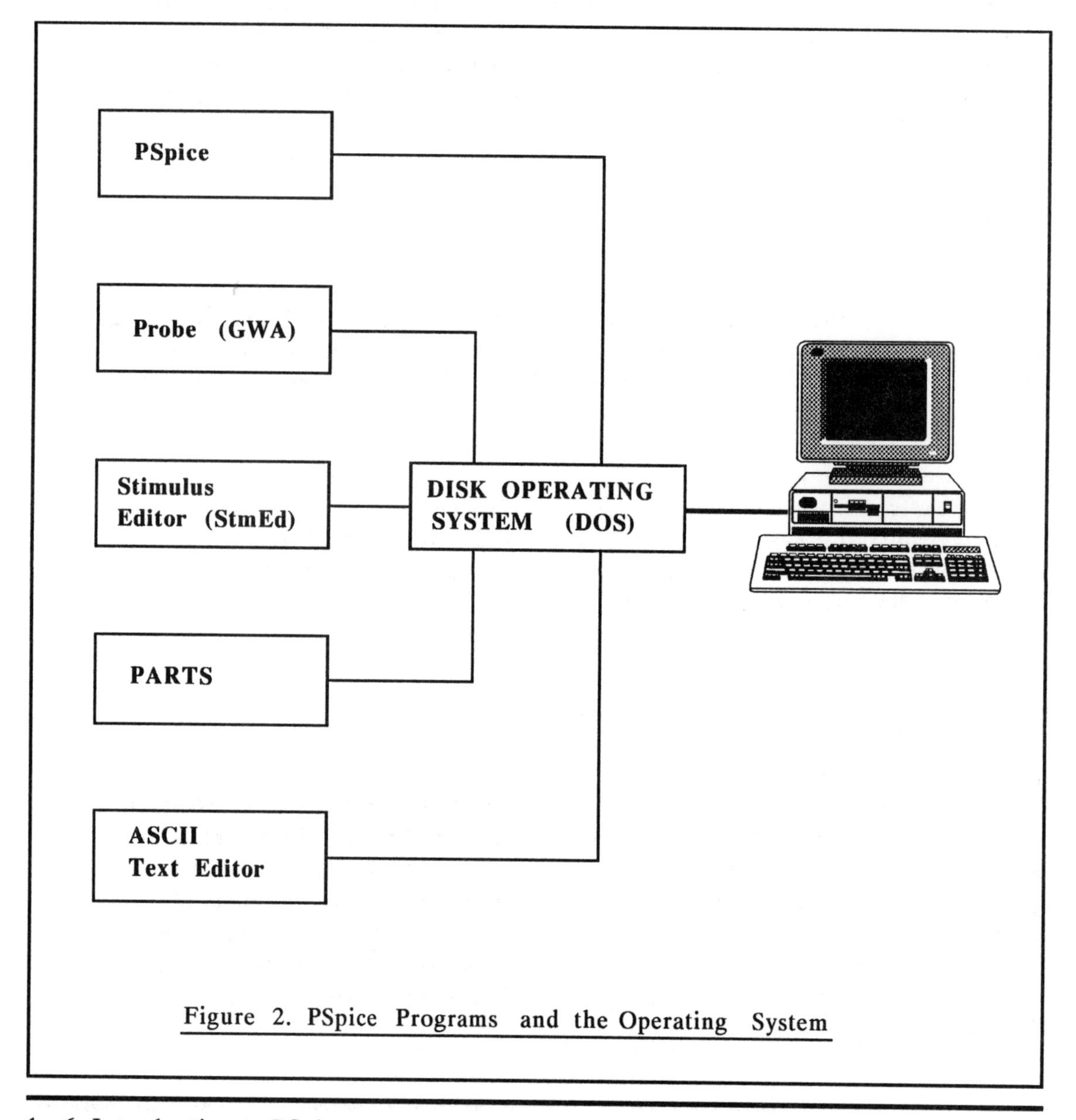

Figure 2. PSpice Programs and the Operating System

The Library

The Device and Subcircuit Library is examined in Chapter 5. The Library program in the full production version of PSpice contains more that 6,000 devices and subcircuits that can be used in circuit design. These devices are modeled from manufacturer's data sheets and are written into ASCII text files. A twenty-five page sample library, EVAL.LIB file, is contained in the evaluation software. It would be a good idea to take a look at the types of data and entry format used in the library by entering "TYPE EVAL.LIB" at the PSEVAL52 directory. Better yet, print a copy.

Standard resistors, inductors (including transformers), and capacitors are not a part of the device library. They are entered into circuits by simply typing R, L, or C, the node locations, and the value of the components. But, in PSpice, even these passive components can be and are modeled as needed. Circuit testing with component tolerance variations is an example of a need to write models.

PARTS: Specialized and New Devices

A tour of the PARTS program is also included in Chapter 5. In PARTS, new devices can be written and modeled and the results saved to user library files. In addition, standard models can be specialized to designer requirements. The PARTS program is sequenced through several interactive screens and, at the last screen, the newly created or modified device is saved to the library. It is also possible to erroneously "modify" and/or mistakenly create an unusable device. For this reason, a copy of PARTS.EXE should be saved in another directory on the hard drive or on a separate diskette. The saved file can be copied back to the PSpice directory if a device becomes inoperative through Parts editing.

The PARTS program is also an independent program within PSpice that can be executed from the operating system. (Figure 2)

Progressive PSpice Text/Manual Experimentation

The experiments detailed in the remaining chapters should reinforce previous learning. Each chapter is a building block in learning PSpice. Initial circuits to be simulated are very basic and later ones grow in complexity and deal with specifics that are encountered in intermediate electronics studies. As with all testing and verification, each detail should be experienced before moving to the next step. Think and work in detail. If you find a challenge, you are learning; if bored, you have peaked the learning curve. Experimentation is the art of discovering detail.

To restate, any similar circuit being studied in the Text/Manual or in your theory textbook can be written into PSpice, simulated, and analyzed. The third objective of the Text/Manual is to provide enough information to you, the student, to complete your own designs. For each circuit studied, it is hoped that at least one, and preferably many, instructor assigned and student designed circuits will be simulated and analysis runs completed to confirm textbook theory.

> The third objective of the Text/Manual is to provide enough information...to complete your own designs.

For our next step in becoming PSpice "programmers", let's take a look at some circuit programming ideas.

III. A Few PSpice Program Notes, Terms, and Definitions

CIRCUIT NODE

Node...can be defined as a connection in a circuit, therefore, a two or more component connection. Think of nodes as solder connections in the circuit. PSpice requires that each node be connected to at least one other node to avoid open circuits. An error message is generated if nodes are not connected. Open circuit simulation is achieved by connecting a large resistor to ground or reference Node 0.

An example of this special treatment of connecting "open" nodes is an output terminal. Output terminals that may not normally have a second connection must have a large resistance connected to Node 0. This resistance would necessarily be large to eliminate adjusting circuit behavior. For example,

```
Rload  12  0  500Meg
```

Nodes are identified with numbers, i.e., 1, 2, 3, etc., or names, IN, OUT, etc. Nodes listed in the above line are 12 and 0. The requirement that each node be numbered or named is a must in PSpice. Node identification and labeling should be a part of the original design before beginning PSpice simulation.

In the first statement below, the current source, I1, is connected at Node 1, the positive node, and Node 0, the negative node. It is a DC source of 1 ampere. Current inside the current source flows from the positive node to the negative node. Current enters the source at the positive node and exits at the negative node. In the second statement, a voltage source of 10 Volts is listed. Node 1 is the positive node and Node 0 is the negative node. This polarity can be reversed by noting the quantity with a minus, for example, -10V in the Vin listing below.

```
I1   1  0  DC  1A
Vin  1  0  DC  10V
```

Node "0" always represents ground, the common reference connection. In PSpice, a DC path must be provided to this common Node 0 for all other nodes in the circuit. Therefore, in designing for PSpice simulation, the "0" node should be established first with all other nodes referencing to it.

Node connections are identified in the circuit in Figure 3.

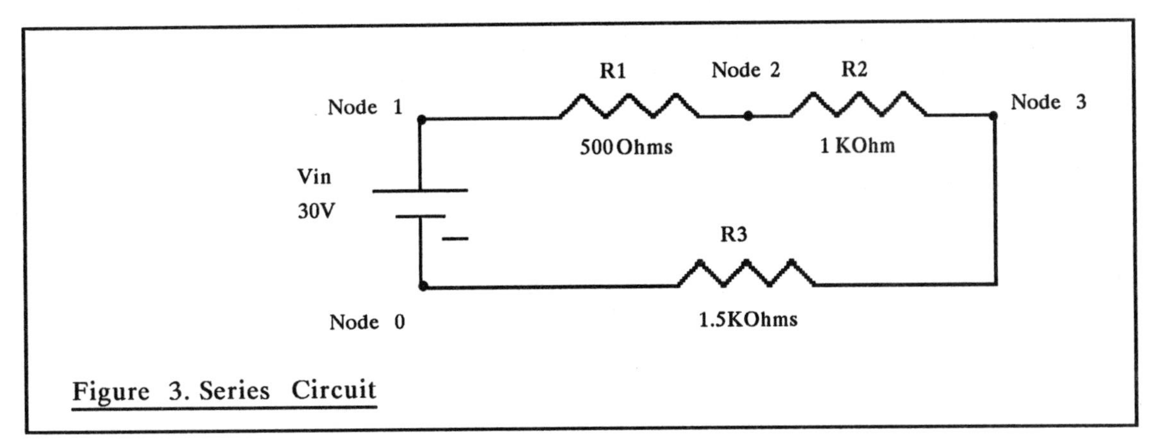

Figure 3. Series Circuit

Writing Nodes

Nodes

| |

Vin is written, Vin 1 0 DC 30Volts

If neither DC nor AC is specified, then Vin is understood to be DC, for example,

Vin 1 0 30V

R1 is written, R1 1 2 .5K, or 500Ohms, or 500

All circuit components must be properly connected, just as if they are soldered into the circuit. Notice that each node is connected to at least two other nodes. The Netlist for this circuit is written on page 10.

NETLIST, The Circuit File (.CIR)

The circuit file, or network list...NETLIST, is the circuit description and analysis program used by PSpice to input the circuit. Circuit files are saved with a .CIR extension. (Backup files are saved with a .CBK extension) The PSpice Control Shell has a Circuit Editor (text editor) and an External Editor for typing in the circuit file. The first line in the netlist is the title line and is always recognized as the title line by PSpice. The second line is the beginning of the program. Usually, description statements begin with the second line and proceed to define the circuit to be simulated.

The netlist also contains control statements that prescribe the test analysis to be performed on the circuit. Control statements also list the outputs required.

In the following example, the source voltage will be swept from 0V to 30V in 1V steps as written in a control statement. .Print, .Plot, and .End are also control statements as listed. .Print defines output requirements and will provide a table of all values of requested voltages and component current during the sweep of Vin. .Print data will be printed in the Output file. .Plot will graph the sweep results of the requested currents and present this graph in the Output file. This data is also available for Probe O'scope analysis and presentation. The .End statement signals the end of the circuit and is a netlist must statement. It is also a must that a period, (.), begin the .End, .Print, and .Plot control statements.

The Series Circuit Netlist: Filename: Chapter1

```
*Chapter 1 Circuit
*Circuit Description statements
Vin  1 0 DC 30Volts                   ;input voltage source
R1   1 2 .5KOhm
R2   2 3 1000Ohms
R3   3 0 1.5K
*Control Statements
.DC Vin 0 30 1                        ;input voltage sweep
.Print DC V(1,2) V(2,3) V(3,0)        ;list swept values in tables
.Print DC I(R1) I(Vin)                ;list swept values in tables
.Plot DC I(R1) I(Vin)                 ;plot swept values
.End
```

Again, in PSpice netlists, the first line is the title line. The asterisk (*) indicates a comment line. And further, a semicolon (;) signals an in-line comment. Comment lines help to keep track of larger circuits and are excellent guides in learning statement meanings. Blank lines can also be inserted for the same purposes. Both types of comment indicators cue the PSpice program to ignore the following information. The asterisk in the title line listed above is not necessary, but is a reminder that the title is not part of the circuit.

The third line in the above netlist begins the circuit description. Control statements usually follow circuit descriptions but the sequence is not critical. PSpice recognizes statements, not locations, but circuit files are easier to read if there is some standard convention and structure. To repeat, the .End statement is a must and the (.) is required.

The Output File

Netlists are converted by PSpice into computer programs. "Run PSpice" analysis of these programs provide ASCII text Output files. The Output file from the analysis of the circuit listed above begins on page 16.

Output files have the same netlist name with an .OUT extension. These files contain circuit analysis data requested in the netlist and other unrequested (default) information. Examples are:

(1) The original input file,
(2) Node voltages, and component currents,
(3) Source current and power calculations,
(4) Data tables and plots specified in .Print and .Plot statements in netlists.

Each section of the file is titled with a separate page banner and report information. We use Browse Output in the Control Shell's Files pop-down menu to observe the data in the file. An Output file is created only when "Run PSpice" is selected in the Control Shell.

Experience and training will be gained by reviewing the Output file in each experiment in the Text/Manual.

How much data will be in the Output file? The report can be very extensive, depending upon the requests and the type of analysis to be performed. If no request for voltage or current has been made, PSpice will default to a complete list of all bias node voltages and source current. This default also includes device currents. The last section of the file contains a job concluded statement, with a listing of total job time of computer use in seconds. Experience and training will be gained by reviewing the Output file in each experiment in the Text/Manual.

The DATA File

PSpice also creates Probe data files (.DAT) during the analysis run. These files are created for data requested in the circuit file control statements. Data files are saved for the Probe Graphical Waveform Analyzer when a .Probe statement is entered into the circuit file or by a default setting.

Data files may be requested in ASCII or binary form using options in the Control Shell.

Circuit Component Value Notations

Specific passive component values are listed as decimal or floating point, i.e., .0005 or 5E-4 which defines 5 x 10 to minus 4.

Standard letter value notations are also acceptable:

K= 10^{3} , MEG= 10^{6} , G= 10^{9} , T= 10^{12}

M= 10^{-3} , U= 10^{-6} , N= 10^{-9} , and P= 10^{-12}

Ohm, volt, farad, and other unit notations are not required in writing circuit files but enhances learning and improves the reading of programs. Examples:

```
R1  1 0 10   is understood  as 10 Ohms
R1  1 0 5K     = 5 KOhms
C1  3 4 2u     = 2 microfarads
L1  6 4 1m     = 1 millihenry
Vin  1 0 30    = 30 Volts, DC
I1  1 0 2m     = 2 milliamps,  DC
I2  1 0 1      = 1 Amp, DC
```

IV. Additional PSpice Particulars

Capacitors in DC Calculations and Coils in .AC and Transient Analysis

Capacitors in DC circuits require special attention in PSpice simulation. Since nodes need a DC path to ground, a capacitor will block this path and will present an open circuit to DC. PSpice will generated an error message; "..Node...is floating." To correct this error, connect a large resistor, 500MEG for example, from that node to Node 0.

Inductors need unique treatment in small-signal analysis (.AC) and transient analysis (.TRAN). Because of self inductance, an inductor is treated as a varying voltage source. This inductor "voltage source" creates a problem if connected across a voltage source in the circuit. As might be expected, current flows unrestricted. Therefore, some resistance must be inserted into the circuit. In short, PSpice treats inductors as voltage sources for .AC analysis and .TRAN analysis, and as shorts or zero voltage sources for DC.

If two inductors are connected in parallel (two voltage sources again), be sure to include a winding resistance in series with each inductor. PSpice will accept this current control.

A Device Simulation and Analysis Note

Model Mathematical Definitions

In the study of physics, we use mathematical equations to define the operation of electronic devices...diodes, BJTs, FETs, Thyristors, and etc. PSpice depends upon these same equations in calculating the various characteristics of these devices. The .MODEL control statement for the device utilizes these equations. Mathemathical models are also written for passive components...re-sistors, capacitors, and inductors, where constant values are not expected.

Control Statements

The following is a list of some PSpice control statements and uses. Some of the statements are not used in the Text/Manual and are included here for introduction purposes. Appendix A also includes these and other statement definitions.

.DC sweep

This statement sweeps the source voltage or source current. At each point in the sweep an analysis of the circuit is performed by calculating the bias points, then the next step is incremented. Typically, .DC sweeps start at one value of the source and calculates all circuit bias points, then moves to the next requested value and recalculates bias points again. An example sweep is used in the circuit in this chapter. Another example is used in drawing collector curves in Chapter 9,

```
.DC  VCE  0  18  .1
```

Sources to be swept with the .DC sweep statement must first be entered into the circuit file. The source voltage, VCE, is listed with a circuit description statement, VCE 2 0, describing VCE at Nodes 2 and 0. This source will be swept in .1V increments, from 0V to 18V. Listed terms are: .DC (type of sweep), VCE (source name), 0 (start voltage value), 18 (stop voltage value), and .1 (increment voltage value).

Types of sweep are linear (LIN), decade (DEC), octave (OCT), and sweep a list (LIST) of values that follow in the statement. Default is the linear sweep.

As the source is swept, measurements of voltages and currents throughout the circuit can be printed and plotted in the .Output file with .Print and .Plot statements. For examples,

```
.Print  DC  V(any node or two terminal device)
.Print  DC  I(any two terminal device)
.Plot  DC  V(specified node or two terminal device)
```

.Print and .plot outputs will include all the calculated sweep step values if the range is not limited. .Print will output to tables and .Plot graphs the output values using keyboard symbols. See the series circuit Output file tables and plots beginning on page 16.

.AC Sweep

The .AC sweep statement provides a small-signal frequency response analysis of the circuit. PSpice will perform three types of frequency definitions in the .AC sweep statement. They are best seen with the examples listed on the next page: DECADE, LINEAR, and LOGARITHMIC sweep.

.AC DEC 100 10K 1MEG. A decade sweep is programmed with 100 points per decade. Decade is a 10X increase in frequency. The range: 10KHz - 1MHz.

.AC LIN 200 10K 1MEG, programs a linear sweep with 200 sweep data points, from 10KHz to 1MHz.

.AC OCT 10 10K 1MEG, programs a logarithmic sweep with the sweep points per octave as the first number in the statement...10. Examples of octave increases in this statement are 10KHz, 20KHz, 40KHz, 80KHz, etc. The sweep range is 10KHz to 1MHz.

.TRAN, Transient Analysis

Circuit response to an input stimulus (input signal) is measured over time with the transient analysis statement, .TRAN. The offset voltage, amplitude, frequency, and other parameters of the input signal are set with a voltage input "generator" statement such as Vin 1 0 SIN(0 .1V 5KHz). This input has an offset voltage of 0V, an amplitude of .1Vpeak, and a frequency of 5KHz. The transient analysis statement specifies the analyses of the circuit with this stimulus during a specific period of time. A typical statement,

```
.TRAN .004ms .4ms 0 .002ms
```

The first entry in the transient analysis line is .004ms. It is the print_step time interval for .print and .plot results from Transient Analysis. This time is also stored for the Probe Graphical Waveform Analyzer program. The print_step time is calculated as final_time divided by 100 for this analysis. (.4ms/100= .004ms)

Final_time is the next entry and is .4ms. It has been set for an analysis and display of two cycles of the 5KHz input. Final_time is calculated as 2(t=1/f) and is the actual run time of the Transient_analysis.

The third entry is results_delay and is set to 0 time. It suppresses the transient analysis start for zero time (no delay). A short delay would allow the circuit to stabilize after the initial response to the incoming stimulus.

Step_ceiling is set to .002ms and is the last entry in the .TRAN statement. The step_ceiling default setting is calculated as final_time/50, which is .4ms/50 = .008ms, and is the maximum step_ceiling time for transient analysis. In this analysis, final_time/200 is used for better Probe graphical displays. (.4ms/200 = .002ms)

Fourier Analysis

Fourier analysis, .FOUR, accompanies a transient analysis, .TRAN, statement. A harmonic decomposition on a stated fundamental frequency, including the

DC, fundamental, 2nd through 9th harmonic components, and harmonic distortion is analyzed with this statement. In the following listing,

.FOUR 500 V(6) ,

a Fourier analysis is requested on a fundamental frequency of 500Hz at Node 6. An example of Fourier analysis is included in the evaluation circuit included in your PSpice software. See Problem 1 of EOC Problems listed below.

Monte Carlo, Sensitivity, and Worst Case Analysis

In Monte Carlo analysis, .MC, component tolerances are randomly changed in multiple runs of .DC, .AC, or .TRAN analyses. The component values used in a particular run are printed along with the results of the analysis. In Monte Carlo all parameters under test are randomly varied and the entire circuit is tested.

In Sensitivity, .SENS, as with Monte Carlo, the first run is made with components at their predicted or "best-case" values. But, unlike Monte Carlo, .SENS analysis checks the circuit with only one component value changed. The output reveals the behavior of the circuit with this one component adjusted. Components tested include resistors, diodes, and BJTs.

Worst Case, .WCASE, is a final sensitivity run with extremes of component tolerance. The user defines the extreme high or low settings.

Conclusion of Chapter 1

Note: Use "The Scientific Method in PSpice Experimentation" (p.iv) when creating exercises and solving problems.

EOC Problems

1. An evaluation circuit is included in your PSpice software, named EXAMPLE1.CIR or EVAL.CIR, depending on your Evaluation Version of PSpice. Print and review the circuit statements and definitions in the circuit file by typing PRINT EXAMPLE1.CIR (or EVAL.CIR) at the PSEVAL52 directory. Print and review the Output file also by typing PRINT EXAMPLE1.OUT (or EVAL.OUT) at the PSEVAL52 directory.

   Review the Output file data and save the file for later references.

2. After completing Chapters 2 and 3, return to Chapter 1 and enter and Run PSpice analysis on the circuit file listed on page 10. Use Probe and Browse Output to view the results of this circuit simulation and analysis. Circuit filename: CHAPTER1

   Study the Output file reports and Probe plots. Compare your Output file with the listing beginning on page 16.

   Note: You will notice a program bug in the I(Vin), (+), table range on page 18.

CHAPTER1.OUT (Output file)

```
****          19:14:03 *********** Evaluation PSpice (July 1992) *****

 *CHAPTER 1 CIRCUIT


 ****     CIRCUIT DESCRIPTION


******************************************************************************

*Circuit Description Statements
Vin 1 0 DC 30Volts
R1 1 2 .5KOhms
R2 2 3 1000Ohms
R3 3 0 1.5K
*Control Statements
.DC Vin 0 30 1
.Print DC V(1,2) V(2,3) V(3,0)
.Print DC I(R1) I(Vin)
.Plot DC I(R1) I(Vin)
.End
```

```
****          19:14:03 *********** Evaluation PSpice (July 1992) *************

 *CHAPTER 1 CIRCUIT


 ****     DC TRANSFER CURVES                TEMPERATURE =   27.000 DEG C


******************************************************************************
```

Vin	V(1,2)	V(2,3)	V(3,0)
0.000E+00	0.000E+00	0.000E+00	0.000E+00
1.000E+00	1.667E-01	3.333E-01	5.000E-01
2.000E+00	3.333E-01	6.667E-01	1.000E+00
3.000E+00	5.000E-01	1.000E+00	1.500E+00
4.000E+00	6.667E-01	1.333E+00	2.000E+00
5.000E+00	8.333E-01	1.667E+00	2.500E+00
6.000E+00	1.000E+00	2.000E+00	3.000E+00
7.000E+00	1.167E+00	2.333E+00	3.500E+00
8.000E+00	1.333E+00	2.667E+00	4.000E+00
9.000E+00	1.500E+00	3.000E+00	4.500E+00
1.000E+01	1.667E+00	3.333E+00	5.000E+00
1.100E+01	1.833E+00	3.667E+00	5.500E+00
1.200E+01	2.000E+00	4.000E+00	6.000E+00
1.300E+01	2.167E+00	4.333E+00	6.500E+00
1.400E+01	2.333E+00	4.667E+00	7.000E+00
1.500E+01	2.500E+00	5.000E+00	7.500E+00
1.600E+01	2.667E+00	5.333E+00	8.000E+00
1.700E+01	2.833E+00	5.667E+00	8.500E+00
1.800E+01	3.000E+00	6.000E+00	9.000E+00
1.900E+01	3.167E+00	6.333E+00	9.500E+00
2.000E+01	3.333E+00	6.667E+00	1.000E+01
2.100E+01	3.500E+00	7.000E+00	1.050E+01
2.200E+01	3.667E+00	7.333E+00	1.100E+01
2.300E+01	3.833E+00	7.667E+00	1.150E+01
2.400E+01	4.000E+00	8.000E+00	1.200E+01
2.500E+01	4.167E+00	8.333E+00	1.250E+01
2.600E+01	4.333E+00	8.667E+00	1.300E+01
2.700E+01	4.500E+00	9.000E+00	1.350E+01
2.800E+01	4.667E+00	9.333E+00	1.400E+01
2.900E+01	4.833E+00	9.667E+00	1.450E+01
3.000E+01	5.000E+00	1.000E+01	1.500E+01

CHAPTER1.OUT (Output file)

```
****          19:14:03 *********** Evaluation PSpice (July 1992) *************
 *CHAPTER 1 CIRCUIT

 ****     DC TRANSFER CURVES                  TEMPERATURE =   27.000 DEG C

******************************************************************************


  Vin          I(R1)        I(Vin)

   0.000E+00    0.000E+00    0.000E+00
   1.000E+00    3.333E-04   -3.333E-04
   2.000E+00    6.667E-04   -6.667E-04
   3.000E+00    1.000E-03   -1.000E-03
   4.000E+00    1.333E-03   -1.333E-03
   5.000E+00    1.667E-03   -1.667E-03
   6.000E+00    2.000E-03   -2.000E-03
   7.000E+00    2.333E-03   -2.333E-03
   8.000E+00    2.667E-03   -2.667E-03
   9.000E+00    3.000E-03   -3.000E-03
   1.000E+01    3.333E-03   -3.333E-03
   1.100E+01    3.667E-03   -3.667E-03
   1.200E+01    4.000E-03   -4.000E-03
   1.300E+01    4.333E-03   -4.333E-03
   1.400E+01    4.667E-03   -4.667E-03
   1.500E+01    5.000E-03   -5.000E-03
   1.600E+01    5.333E-03   -5.333E-03
   1.700E+01    5.667E-03   -5.667E-03
   1.800E+01    6.000E-03   -6.000E-03
   1.900E+01    6.333E-03   -6.333E-03
   2.000E+01    6.667E-03   -6.667E-03
   2.100E+01    7.000E-03   -7.000E-03
   2.200E+01    7.333E-03   -7.333E-03
   2.300E+01    7.667E-03   -7.667E-03
   2.400E+01    8.000E-03   -8.000E-03
   2.500E+01    8.333E-03   -8.333E-03
   2.600E+01    8.667E-03   -8.667E-03
   2.700E+01    9.000E-03   -9.000E-03
   2.800E+01    9.333E-03   -9.333E-03
   2.900E+01    9.667E-03   -9.667E-03
   3.000E+01    1.000E-02   -1.000E-02
```

CHAPTER1.OUT (Output file)

```
****          19:14:03 *********** Evaluation PSpice (July 1992) *************

 *CHAPTER 1 CIRCUIT

 ****      DC TRANSFER CURVES                 TEMPERATURE =   27.000 DEG C

******************************************************************************



 LEGEND:

*: I(R1)
+: I(Vin)

  Vin           I(R1)
(*)----------    -5.0000E-03    0.0000E+00    5.0000E-03    1.0000E-02    1.5000E-02
(+)----------    -1.5000E-02   -1.0000E-02   -5.0000E-03    8.6736E-19    5.0000E-03
                        - - - - - - - - - - - - - - - - - - - - - - - - - -
  0.000E+00  0.000E+00 .           *            .           +            .
  1.000E+00  3.333E-04 .           .*           .          +.            .
  2.000E+00  6.667E-04 .           . *          .         + .            .
  3.000E+00  1.000E-03 .           .  *         .        +  .            .
  4.000E+00  1.333E-03 .           .  *         .        +  .            .
  5.000E+00  1.667E-03 .           .   *        .       +   .            .
  6.000E+00  2.000E-03 .           .    *       .      +    .            .
  7.000E+00  2.333E-03 .           .     *      .     +     .            .
  8.000E+00  2.667E-03 .           .      *     .    +      .            .
  9.000E+00  3.000E-03 .           .       *    .   +       .            .
  1.000E+01  3.333E-03 .           .        *   .  +        .            .
  1.100E+01  3.667E-03 .           .         *  . +         .            .
  1.200E+01  4.000E-03 .           .         *  . +         .            .
  1.300E+01  4.333E-03 .           .          * .+          .            .
  1.400E+01  4.667E-03 .           .           *.+          .            .
  1.500E+01  5.000E-03 .           .            X           .            .
  1.600E+01  5.333E-03 .           .           +.*          .            .
  1.700E+01  5.667E-03 .           .          + .*          .            .
  1.800E+01  6.000E-03 .           .         +  . *         .            .
  1.900E+01  6.333E-03 .           .         +  . *         .            .
  2.000E+01  6.667E-03 .           .        +   .  *        .            .
  2.100E+01  7.000E-03 .           .       +    .   *       .            .
  2.200E+01  7.333E-03 .           .      +     .    *      .            .
  2.300E+01  7.667E-03 .           .     +      .     *     .            .
  2.400E+01  8.000E-03 .           .    +       .      *    .            .
  2.500E+01  8.333E-03 .           .   +        .       *   .            .
  2.600E+01  8.667E-03 .           .  +         .        *  .            .
  2.700E+01  9.000E-03 .           .  +         .        *  .            .
  2.800E+01  9.333E-03 .           . +          .         * .            .
  2.900E+01  9.667E-03 .           .+           .          *.            .
  3.000E+01  1.000E-02 .           +            .           *            .
                        - - - - - - - - - - - - - - - - - - - - - - - - - -



          JOB CONCLUDED

          TOTAL JOB TIME            13.84
```

Chapter 2

PSPICE CONTROL SHELL

Objectives: (1) To introduce the Control Shell
(2) To write a small circuit file to enable accessing the Control Shell
(3) To introduce the Pop-Down Menus: Files, Circuit, StmEd, Analysis, Display, Probe, and Quit.

Note: A short wait in some parts of the program may be necessary if a coprocessor is not installed. Avoid hitting the Enter key more than once during these delays. A second Enter will be executed next and you may move to an unfamiliar screen. Double clicking with the Mouse should also be avoided for the same reason.

The following is a survey of the PSpice Control Shell, the main menu in PSpice. Each step introduces a different selection in the menu and F1=Help statements are available at most locations. The chapter can be repeated as a tutorial exercise to improve proficiency in operating the program.

Control Shell is a menu driven supervisory program for the various programs in PSpice. It provides a platform from which these programs are easily accessed and coordinated. For examples, Probe and StmEd are independent PSpice programs that can be operated from the Control Shell. The easy access and return provided, in addition to being a "home" screen for the operator, makes the PSpice Control Shell uniquely user friendly. Let's get started.
Type PS at the PSEVAL52 directory. Press Enter.

The Control Shell displays a titleblock and two menu bars at the bottom of the screen.

```
                    PSpice Control Shell - ver 5.2
Files     Circuit     StmEd     Analysis     Display     Probe     Quit




        Current File:____________________________        (New)
F1=Help  F2=Move  F3=Manual  F4=Choices  F5=Calc  F6=Errors  Esc=Cancel
```

MENUS AVAILABLE: Files, Circuit, StmEd, Analysis, Display,
Probe, and Quit.

Notice that Files, Probe, and Quit Menus are highlighted titles and are the only selectable Menus at this point.

I. Files Pop-Down Menu

Notes: Press "F1=Help" at each of the following selections for additional information where available. Press Esc to exit Help.
The Control Shell pop-down menus should display automatically after a file is selected. If not, press Enter at each selection.

Take a look at Files in the Control Shell.
Select Files, and Enter, to display the Files pop-down menu.

Since we have not selected a file to edit, only two of the Files selections are available...Display/Prn Setup and Current File.

Select Display/Prn Setup, and Enter.

The "Configuring: pspice.dev" menu appears. The selections chosen when PSpice was installed are displayed. We may type in new Display, Port, and Printer configurations. The program returns to the Control Shell when we have selected the Printer. Press "Esc" to return to the Control Shell.

Select "Current File...", and Enter.

A "Define Input File" window with a request that a "Current File Name:" be entered. A previously named file may be called up or a new circuit file name may be entered. We will return to this window later. For now, we will take a quick look at Probe and Quit in the Control Shell. We will also examine the Function keys and review a mouse note.
Press Esc to return to the Control Shell.

II. Probe Pop-Down Menu

Probe, or Probe Graphical Waveform Analyzer, is the "oscilloscope" of PSpice. A data file, with a .DAT extension, is necessary for Probe O'scope observations. Data files are created during PSpice analysis of the circuit. A default setting or .probe statement in the circuit file will save data files for Probe use.
Select Probe, and press Enter if the Probe pop-down menu is not displayed.
Select Format, and Enter.

By entering "B" or "T" in this window we may select the format Probe will use in generating Probe files...binary or ascii.

Press "B" for binary, and Enter.

Next, run through the remaining selections and note results.

Select:

(1) Run Probe, and Enter.
    Run Probe requires a "Probe File Name:".
    Press F4 and a list of all data files, .DAT, will be displayed.
    Press Esc twice to return to the Probe pop-down menu.
(2) Log to File..., and Enter.
    Create log file? (Y/N) Press "N" for No, or press Esc to exit.
(3) Command File..., and Enter.
    Use command file? (Y/N) Press "N" for No, or press Esc to exit.
(4) Auto-run..., and Enter.
    Automatically run Probe after analysis? (Y/N)
    By selecting "Y", all PSpice runs will be automatically followed by a Probe run. Press "Y" for Yes and exit to the Control Shell.

III. Quit Pop-Down Menu

Select Quit in the title bar, and the Quit menu is displayed.

The options are "Exit to Dos" and "Dos Command..." Here, we may quit PSpice altogether by selecting the "Exit to Dos" highlight. We may also select "DOS Command..." and use DOS commands without leaving PSpice. Select "DOS Command..." and Enter. Type DIR/P at the "DOS Command" line and, using the Enter key, page through the directory files. The last Enter should return you to the Control Shell.

Function Keys

The following is a list of function keys with short definitions. Some keys are not operational and data is limited in all keys in the evaluation version of PSpice. Press Esc to exit function key selections.

F1=Help	Provides help on the area you are currently working on.
F2=Move	Move or resize pop-up windows.
F3=Manual	Calls up the On-Line Manual. For example, in Editor, press F3 and you may select a manual page that provides editing notes.
F4=Choices	Provides a list of choices where available.
F5=Calc	Provides a calculator. (Not fully operational)
F6=Errors	Provides a list of errors in the current file.

Mouse Notes

Moving the mouse is the same as using the Arrow keys. Clicking the left button is the same as Enter, and clicking both buttons is Esc.

Try these mouse commands by moving around the titleblock.

IV. Files: Entering a New File

The remaining selections in the PSpice Control Shell will be available to explore when we define a current file.

At the Files title, select Current File... and Enter.
At "Define Input File", type **BJTAMPLIFIER** and press Enter.
Select Files, Enter, Edit, Enter.

The "Circuit Editor" window is opened for you to write a circuit "netlist" using description and control statements. This circuit file will be simulated and analyzed by PSpice.

Enter the file as listed below.

```
                    PSpice Control Shell - ver 5.2
                 Circuit Editor line: 1 col: 1 [Insert]

*BJT AMPLIFIER; Circuit Description statements
VCC 6 0 DC 20V
Vs 1 0 AC 1 SIN(0 .5V 1KHz)          ;input
Cin 1 2 8u                                          ;(Sketch circuit here)
R1 6 2 118K
R2 2 0 18K
R3 6 4 8.5K
Q1 4 2 3 bjt ;collector, base, emitter
R4 3 0 1.8K
Cout 4 5 8u
R5 5 0 1.5Meg          ;output
*Control statements
.model bjt npn(bf=70 cjc=4p rb=90)
.probe
.ac dec 5 10m 100Meg
.tran .02ms 2ms 0 .01ms
.end

              Current File: BJTAMPLIFIER.cir      (NEW)
F1=Help  F2=Move  F3=Manual  F4=Choices  F5=Calc  F6=Errors  Esc=Cancel
```

Note: Recheck your netlist. If your file is different from the one written above, use the Arrow keys and Insert/Overtype to correct your circuit file.

Study the terms and arrangement of individual statements, then press Esc. The program now prompts to,

"Save changes (to a temporary working file) or Discard" (S/D)? (S)

Press "S" or Enter to save the new file. This brings you back to the Files selection in the Control Shell.

ERRORS!

PSpice will detect errors in the input file and prompt you. You will receive this prompt when you save the new circuit file or, later, when you attempt to "Run PSpice" analysis.

There are two ways to view a list of circuit file line errors. One way is to press F6 as was noted earlier. The other is to,

Select Circuit in the Control Shell, and Enter.

Select Errors, and Enter. The Errors selection is available only if errors exist.

Follow prompts to edit the circuit file.

You will notice that other menus in the Control Shell have opened after the circuit file has been saved without errors.

Circuit Diagram

Sketch the BJTAMPLIFIER circuit to the right of the circuit file on page 2-4. Mark node locations on your drawing.

Run PSpice

We will skip ahead in our tour of the PSpice Control Shell and "Run PSpice" on the BJTAMPLIFIER circuit. The circuit will be simulated and analyzed. Remember, this is where circuit file errors may also be detected.

Move across the top of the menu to "Analysis" and select "Run PSpice." Press Enter. If all goes well the Probe title screen will be displayed when circuit analysis is completed. At Probe, select "Exit_program," and Enter, to return to the PSpice Control Shell. Great job!

PSpice has created an Output file, BJTAMPLI.OUT, that we will view next with the Browse Output selection. In addition, a data file, BJTAMPLI.DAT, was created to be used by the Probe Graphical Waveform Analyzer.

At this point, you have written a circuit file, BJTAMPLIFIER.CIR, and PSpice has simulated and analyzed that circuit. The entire program is now operational.

Select Files, and Enter.

Notice that all selections in the Files pop-down menu are now available (highlighted).

V. Files: Browser Output and Other Selections

Select Browse Output, and Enter.
The Output Browser window is displayed.

Use PgDn, PgUp, and Arrow keys to page through the Output file, BJTAMPLI.OUT. Study the various data as you scan the pages.

First screen :	Circuit Title and Circuit Description
Second screen:	Transistor model parameters are displayed
Third screen :	Small Signal Bias Solution page with Temperature. Default temperature is 27 DEG. C.
Fourth screen:	Initial Transient Solution page with Temperature

Voltage Source Currents and Total Job Time conclude the Output file.

Press Esc to return to the Control Shell.

Other Files Selections:

Save File	Press Enter and the Current File, BJTAMPLIFIER is saved to disk. Follow prompts by pressing Enter.
X-Ext.Editor	Provides an edlin editor. Press Enter. At the edlin prompt (*), type "E", and Enter, to return to the Control Shell. (See Edlin Editor below)
R-Ext.Browser	Provides the same paging as "Browse Output" on a full screen. Press Enter where necessary. You are temporarily out of the Control Shell while using the R_Ext. Browser. Press Enter for paging and automatic return to to the Control Shell.

Edlin is a line-oriented text editor that can be used as the PSpice X - External Editor. The currently selected file is edited or a new file may be created. When External Editor is selected, the original circuit file is saved to a backup (.CBK) file and the edlin prompt (*) is displayed.

Note: The X-External Editor in the PSpice evaluation version is not a fully operational text editor. Edlin commands are limited.

Select X - External Editor, and Enter.
The edlin prompt (*) should be displayed.

Edlin Editor Cont.

To list consecutive lines:
Type L, and Enter, for a numbered line list of the circuit to be edited.

To Edit:
Type the number of the line to be edited, and Enter.
Type changes to be made in the line. (Use the right arrow key to move along the line to the point to be edited)
Press Enter after editing the line.
Type L, and Enter, for a new list of the edited file.

To end and exit the X - External Editor:
Type E at the Edlin prompt (*), and Enter.
The program returns to the Control Shell.

Entering a new file using X-External Editor:
A new circuit file can be entered with the External Editor.
Select Files, Enter, Current File, Enter, and name the new file.
Select X - External Editor and Enter and the edlin prompt (*) appears with the note "New file."
Type I (for Insert), and Enter, to begin typing the new lines.
Type your new circuit netlist and press Enter after each line.
Press Control-C after entering the .End statement.
Type L, and Enter, for a list of the new file lines.
Type E, and Enter, to exit the external editing mode.

Take a look at the new file by selecting Files-Edit at the Control Shell. Your file should appear exactly as you typed, without the numbered lines. Editing is much easier and faster at the Circuit Editor but X-External Editor is another option available in PSpice. Press Esc, and "D" for Discard.

Recall the BJTAMPLIFIER circuit for the next exercise:
Select Files, Current File, and Enter.
At Circuit File Name: type BJTAMPLIFIER, and Enter.

VI. Circuit Pop-Down Menu

Select Circuit in the Control Shell, and Enter. Notice that the "Errors" selection is not available (not highlighted). There are no errors in the writing of the current file...BJTAMPLIFIER. (Correct any errors that have been inadvertently introduced into the circuit file)

Devices
Select "Devices...", and Enter.
The "Select Device to Change" window is displayed.

Circuit component changes can be made at this window. Use the Arrow keys to move through the selections. For example,

Arrow to R4= 1.800K, and Enter.
At the R4 window, select "resistance= 1.800K", and Enter.
At "new value?", type 1750, and Enter.

The new value of R4 is displayed and is also edited into the circuit file. Next, examine this editing by returning to the Circuit Editor.
Press Esc twice to return to the Control Shell.
Select Files, Edit, and Enter. Check R4 for the new size, 1.750K.
Press Esc, and Save.

Models

At the Circuit pop-down menu, select Models, and Enter.
The "Models" window appears with the bjt npn model listed. Press Enter for a list of model parameters. These and other parameter quantities define the NPN transistor model being used in the circuit.

Edit **bf, maximum forward beta:**
Arrow to "bf=70", and Enter.
At "new value?", type 75, and Enter.
Press Esc twice to exit the models windows.

Return to the Circuit Editor and check program editing of bf=75.
Press Esc, and save.

See Appendix B for a list of BJT model parameters.

Select Circuit in the Control Shell, and Enter.

Parameters

Select Parameters, and Enter.
The "Global Parameters" screen is displayed. These global parameters apply to the entire circuit or subcircuit under examination. For example, Gmin is the minimum conductance used for any branch in the circuit.
Press "Esc" to return to the Control Shell.

This completes the Circuit menu. We will now examine the StmEd program and take a look at editing the input signal.

VII. StmEd -- Stimulus Editor Pop-Down Menu

StmEd

Select StmEd, and press Enter.
Select Edit, and Enter.
An inquiry window may be displayed. If so, press "S" or Enter to save any changes. Follow prompts and continue to press Enter as neeeded.

Note: The StmEd program will be loaded and will boot up to the input signal (stimulus) of the current circuit file...BJTAMPLIFIER.

"WOW!" "Impressive, isn't it?" (Sketch graph and waveform on page 2-16.)

The Screen displays the "input stimulus" or transient analysis input signal that was programmed at the Circuit Editor window. Do you remember the input values in the BJTAMPLIFIER netlist?

Vin 1 0 AC 1 SIN(0 .5V 1KHz)

(AC 1 is the input for AC sweep small-signal analysis)
A quick check of the waveform shows a sine wave, .5Vpeak, recurring at a 1ms time period. Using f = 1/t, the frequency is indeed 1KHz.
The input signal can be edited at this screen. It is called stimulus editing, and the program used to edit is called StmEd, for short. We will return to the StmEd display later, but for now, study the selections at the bottom of the screen and select Exit, and Enter.
The display should be the Stimulus Editor title screen with some MicroSim Corporation messages:

|Stimulus Editor|
Stimulus Editor for PSpice
Version 5.2 - July 1992
(C) Copyright 1985 - 1991 by MicroSim Corporation

Exit_program **Start_editor** **Abort_program**

Select "Start_editor", and Enter.

You went back to that impressive StmEd display, didn't you? You are running the StmEd program.

Select Exit, then Enter.

At the Stimulus Editor title screen, select "Exit_program", and Enter. The Control Shell should be displayed.

Nice going!

We will now finish the StmEd pop-down menu. Select StmEd, and Enter.

Select and press Enter:

Command File... Use Command file (Y/N)?

Pressing "Y" enables the use of a command file, CMD, with the Stimulus Editor. Press "N" or Esc to exit.

Log to File... Create log file (Y/N)?

Press "N" or Esc to exit.

VIII. Analysis Pop-Down Menu

In an earlier section, we put PSpice to work analyzing the circuit file, BJTAMPLIFIER.CIR. We will again select "Run PSpice" when we finish the tour of the Analysis pop-down menu. But first, let's take a look at the other options in this menu. Again, all selections are available since PSpice has analyzed the current file.

Select Analysis, and Enter.

Press F1=Help key at each selection. Press Esc to return to the Analysis pop-down menu.

Pressing Esc accepts the default setting.

Select, and press Enter:

AC & Noise... "AC Analysis" window with Sweep Type options. The Type, Start, End, and Pts/Decade should reflect line 15 of the BJTAMPLI.CIR file. A decade sweep is requested, beginning at .01Hz, ending at 100MHz, and 5pts/decade. AC analysis parameters remain unchanged whether analysis is Enabled or not. ("Y" should be indicated)

Press Esc to exit.

Select, and press Enter:

DC Sweep...	"Main DC Sweep" window, Enable (Y/N) Press "N" or Esc to exit.
Transient...	"Transient Analysis" window. Enable Transient and/or Fourier analysis. See .tran control statement in file. Note final time= 2ms, print step= 20us, and step ceiling= 10us. Press Esc.
Parametric...	"Parametric Sweep" window. Allows enable/disable of parametric sweep. The sweep will be performed on one variable for all analyses of the circuit. Press Esc.
Specify Temperature...	"Temperature" window is displayed. Circuit ambient temperature can be specified in degrees Celsius. Default is 27 degrees C. Press Esc.
Monte Carlo...	"Monte Carlo" window allows enable/disable of Monte Carlo analysis. Selecting "Y" will expand the window to specify parameters. (Monte Carlo allows the assignment of tolerances to component values which are then used to make multiple runs in DC, AC, or Transient analysis with variations in those values. In this analysis, all values are varied randomly each run. Coupled with Sensitivity and Worst Case analysis, a complete picture of the various results can be viewed. In these last two analyses, only one value is varied each run by a fixed percent. Sensitivity examines circuit behavior changes for each component value. Worst Case will produce the worst case waveform for high or low value) Press Esc twice.
Change Options...	"Options" window. Change options to be used in the simulation/analysis.

Look over the analysis options that are available to be adjusted. To understand these options we must study each one individually, including abbreviations and value ranges for each. Select several options and browse using the Up/Down Arrow keys. Press F1=Help at each option. Press Esc to exit Help.

At **"nopage"**, press "Y", and continue to use the Down Arrow key until you reach the last selection in the table. Press Enter to return to the Analysis Pop-Down menu. You have entered a "nopage" statement into the circuit file that will suppress paging and banners in the Output file. Return to the Circuit File and check this new control line. Return to the Analysis pop-down menu.

We are ready to put PSpice to work again.

RUN PSPICE...

Select Run PSpice, and Enter.

Again, this is the analysis run as PSpice executes the control statements listed in the BJTAMPLIFIER circuit file. The circuit is being simulated and analyzed by the program and the results are being written to Output (.OUT) and data (.DAT) files. With data files, transient analysis and AC sweep outputs are stored to be used by the Probe Graphical Waveform Analyzer.

The Probe program will automatically run since it was specified in the original circuit statements as ".probe". We also selected an automatic Probe run by selecting "Y" in the Probe pop-down menu.

The Probe title screen will be displayed when the analysis is completed.

| Probe |
Waveform Analyzer for PSpice
Version 5.2 - July 1992
(C) Copyright 1985 - 1992 by MicroSim Corporation

Circuit: BJTAMPLIFIER **Temperature:** 27.0
Date/Time run:

Exit_program **AC_sweep** **Transient_analysis**

Select AC_sweep, and Enter.

A Probe graph is displayed with frequency (10mH-100MHz) as the X_axis. The following steps will display the output voltage vs frequency.

Select Add_trace, and Enter.

Press F4 for a selection of circuit variables to be plotted.

Arrow to V(5), the circuit output, and press Enter to make the selection. Press Enter again to display V(5).

The screen trace is the familiar output curve of a swept capacitor coupled amplifier. AC_sweep input, AC 1 in the stimulus statement, is 1 Volt or unity input. AC_sweep measurements will be discussed later. The Y_axis voltage range was added automatically. The circuit input frequency scans from ~Dc to above 100MHz. The lower critical frequency is determined by the capacitive coupling limitations and the upper critical frequency is controlled by several factors including transistor internal capacitance.

(Sketch graph and plot on page 2-16)

The simulation by PSpice and the Probe display of actual circuit operation is outstanding. The frequency sweep range and analysis were specified in the circuit control statement,

.ac dec 5 10m 100meg

Defined as:

.ac	Calculate small-signal response over a range of frequencies
dec	sweep type (Linear/Octave/Decade)
5	number of sweep points per decade
10m	start frequency, 10mHz, or .01Hz.
100meg	end frequency, 100MHz

Exit AC_sweep by selecting Exit, and Enter. The Probe title screen is again displayed.

Using Probe, we will now examine transient analysis input and output voltages and waveforms of the BJTAMPLIFIER circuit.

Transient Analysis

Select Transient_analysis, and Enter.

Time, in milliseconds (0s - 2ms), is displayed on the X_axis.

Set the Y_axis range:

Select "Y_axis," and Enter.

Select "Set_Range," and Enter.

At Enter a Range: type -3V 3V , and Enter.

Accept Exit, and Enter, to leave the Y_axis menu and return to the original transient analysis graph. Note the new Y_axis values.

The Input Stimulus

Select Add_trace, press F4 and arrow to V(1), the input voltage, and Enter, and Enter again to plot the input signal. The input stimulus specifications were stated in the circuit statement,

SIN(0 .5V 1KHz)

The input is displayed as Vpeak= .5V.

Output Signal

Repeat the above Add_trace steps for V(5), the output voltage. The output is approximately 2.3Vpeak. Gain= Vout/Vin = 4.6

The graph displays input vs output voltages in this transient analysis. Probe is presenting a graph of the operation of our circuit from which circuit adjustments may be made...component sizes, input signal, VCC, and etc. In short, we are looking at variable experimental data without building the circuit or turning on test equipment. PSpice is on the job.

The point to grasp is that this is computer circuit simulation and analysis and the results of PSpice "breadboarding" and testing are displayed on the screen. The inverted output is observed. Voltage gain is easily determined. Even circuit Class A operation can be analyzed with a few additional measurements. Study the data with the original circuit in mind. Remember, PSpice is an analytical tool to help the designer verify and improve designs. But, what a tool! (Sketch the graph and both input and output waveforms on page 2-17)

Select Exit, and Enter.

At the Probe title screen, select "Exit_program", and Enter. The PSpice Control Shell is displayed.

We have looked at only a few of the options that are available in Probe menu selections. In AC_sweep and Transient_analysis we were introduced to only initial menu screens. We will return to Probe in a later chapter, but for now, let's complete the Control Shell menus.

IX. Display Pop-Down Menu

Select Display, and Enter.

Select and Enter:

Print	Prints specified output variables (.print), i.e., Analysis Type (Ac/Dc/Tran/Noise). A single analysis type is output to print tables. A list of variables are then specified and values are printed for each output variable as they change value. Press Esc to exit.

Let's review the Probe and Quit Menus before finishing this chapter.

X. Probe Pop-Down Menu

Select Probe, and press Enter.

Select and Enter at the following selections:

Run Probe	Runs Probe on the current circuit if PSpice analysis has been completed. Other data files may be specified if a current file has not been named. Select Exit_program to exit.
Auto-run...	Automatically runs Probe after PSpice analysis, Options (Y/N)? Press "Y" for Yes, or Esc to exit.
None/Some/All...	Saves output variables for Probe: Options: All/Some/None Press "A" for All.
Command File...	Use Command File (Y/N)? Press "N", or Esc to exit.
Log to File...	Create Log File (Y/N)? Press "N", or Esc to exit.
Format...	Specify Probe file format; Binary or Text Binary file = .dat Text file = .txt Binary is the format generated by .probe. Text is the format generated by .Probe/CSDF. Press "B" for Binary, or Esc to exit. Press Esc again.

XI. Quit Pop-Down Menu

Select Quit, and Enter.

Again, the options are "Exit to Dos" and "Dos Command..." Here, we may quit PSpice by selecting the "Exit to Dos" highlight.

"Dos Command..." provides a window to allow DOS commands to be entered and, after command execution, returns to the Control Shell. For examples: (1) to scan the library (EVAL.LIB), (2) to Run Probe or StmEd from DOS, or (3) to take a look at a list of files in this program. We will look at the first example and scroll through the 25 pages of the evaluation library.

Select Dos Command..., and Enter.

At Enter Command..., type "type EVAL.LIB", and Enter at prompts.

Press pause to stop scrolling. Press any key to continue. The program will return to the Control Shell.

Select Quit, and Enter.

Select Exit to Dos, Enter, and follow prompts.

You are at the Dos prompt. Great job!

This completes the survey of the PSpice Control Shell. Repetition will reinforce learning these basic moves within this exciting circuit simulation and analysis program...PSpice.

Conclusion of Chapter 2

EOC Exercises

Stimulus Editor

Sketch the graph and input stimulus from page 2-9.

AC_sweep

Sketch the graph and AC_sweep response from page 2-13.

Transient_analysis

Sketch the graph and both input and output waveforms from page 2-14.

Notes:

Chapter 3

PROBE "OSCILLOSCOPE"

.DC SWEEP

Objectives: (1) To introduce the Probe Graphical Waveform Analyzer program
(2) To introduce the .DC sweep control statement

Note: A short wait in some parts of the program may be necessary if a coprocessor is not installed. Avoid hitting the Enter key more than once during these delays. A second Enter will be executed next and you may move to an unfamiliar screen. Double clicking with the Mouse should also be avoided for the same reason.

PROBE, or Probe Graphical Waveform Analyzer, is an independent program that is part of PSpice. PROBE.EXE can be entered from the PSpice Control Shell, as we learned in Chapter 2, or directly from the DOS prompt. This program provides an "O'scope" view of various voltage and current measurements including inputs and outputs of the circuit under test. Analysis is possible only with circuit data files (.DAT) that have been created by a PSpice run. We will examine Probe in this chapter, but first, let's take a look at the DC Sweep control statement that will be needed when we operate Probe.

I. .DC Sweep - An Introduction

The benefit of being able to sweep DC inputs and read voltage and current changes throughout the circuit is another analysis capability in PSpice. DC Sweep simulation of a circuit's operation can be demonstrated in a typical voltage divider. (Figure 1) In this circuit, Vpwr, the voltage source, will be swept from 0V to 12V to provide the sweep variable. All voltages and currents in the circuit will, of course, change linearly. .Print and .Plot statements are used to make several measurements during the sweep. Ohm's Law analysis of this circuit will provide data that will be recorded in tables and plots in the Output file, DCSWEEP.OUT. Browse Output will be used to view this data. Probe will use the data file, DCSWEEP.DAT, for more detailed graphic observations.

Type PS at the PSEVAL52 directory, and Enter.

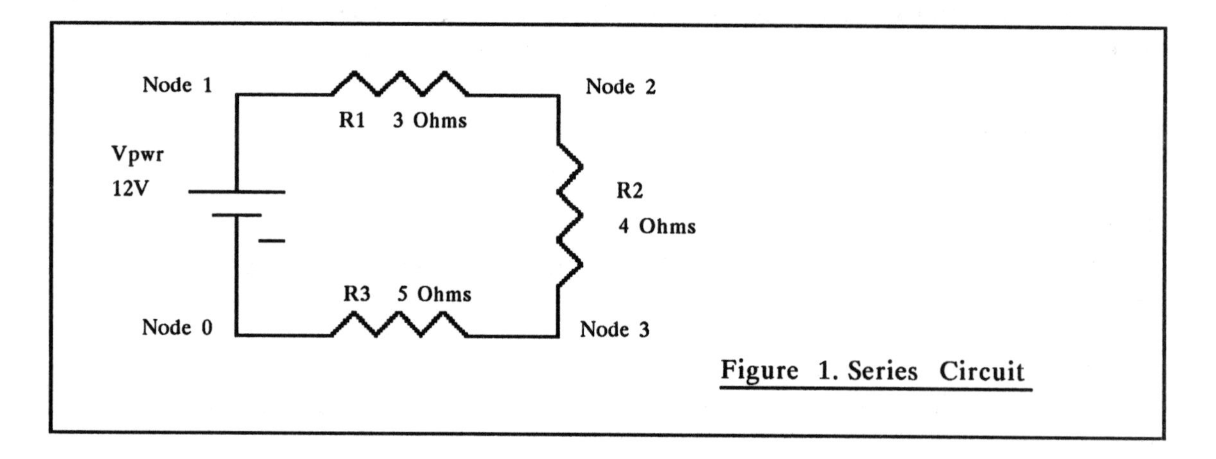

Figure 1. Series Circuit

Select Files, Enter, Current File, and Enter.
At Current File name: type DCSWEEP, and Enter.
Select Edit, and Enter.
At the Circuit Editor, type the circuit netlist. Include the comment statements...lines preceded by an asterisk, (*), or a semicolon, (;).

The Netlist

```
*DCSWEEP...AN INTRO TO .DC SWEEP
Vpwr 1 0 DC 12V
R1   1 2 3Ohms          ; 3 ohms
R2   2 3 4Ohms
R3   3 0 5Ohms
*Control statements
.PRINT DC I(Vpwr) V(1) V(2) V(3) V(2,3)  ; print tables of
.WIDTH OUT=80                            ; these values
.TEMP 30
.PROBE
.DC Vpwr 0 12 1         ;sweep Vpwr 0V - 12V in 1V steps
.PLOT DC V(2)           ;provide a plot of V(2) vs Vpwr
.PLOT DC V(2,3)         ;plot voltage drop across R2
.PLOT DC I(R1)          ;plot current through R1
.END
```

Press Esc, then "S", or Enter, to save the circuit file.
Calculate the voltage and current ranges expected as Vpwr is swept from 0V to 12V.

Run PSpice

Select Analysis, Enter, Run PSpice, and Enter.

The initial Probe menu screen will be displayed when the PSpice run is completed. Select Exit as needed and return to the Control Shell.

Control Statements

.PRINT

Prints the requested values in the Output file. Browse Output is used to view the data.

.WIDTH OUT=80

Sets the width of the Output file to 80 columns. Two column settings are available, 80 and 132. If unset, default is 80.

.TEMP 30

Sets the temperature for the run at 30 degrees C. Default setting is 27 degrees C. Temperature can be set at any test value, or swept in a DC sweep statement. For example, voltage and current sources and individual component values can be swept while stepping temperature levels.

.PROBE

Saves all voltage and current values for Probe. We will be using plots of these quantities in the Probe program.

.DC Vpwr 0 12 1

Sweeps the input voltage from 0V to 12V in 1 Volt steps. Three sweep types can be specified, Linear (LIN), Decade (DEC), and Octave (OCT), and each will give a corresponding plot. The default sweep is Linear and is used in the analysis of this example circuit.

.PLOT

Graphs any type sweep and the swept values are presented on the X_axis with the input sweep values (Vpwr) on the Y_axis. Plots are presented in the Output file along with .Print data tables and are examined using Browse Output. The various values requested could have been combined on a single plot, but are more easily read on individual graphs. It is a good idea to limit the number of plot points, thereby limiting the size of graphs.

A shorter range of values may be specified on plots. If the range is not entered, default is the full range of swept values.

The graphed plots use keyboard symbols.

BROWSE OUTPUT

Select Files, and Enter.
Select Browse Output, and Enter, to view the Output file.
Use PgUp, PgDn, and the Arrow keys to move around in the file.

While browsing the Output file, examine the voltages and current specified in these circuit description and control statements:

```
Vpwr 1 0 DC 12V
.PRINT
.PLOT
.DC Vpwr 0 12 1
```

The Output File

First screen: DCSWEEP Circuit Title and Circuit Description

Second screen: Temperature-Adjusted Values Temperature= 30 DEG C

Third screen: DC Transfer Curves Tables (.Print command in circuit file)

Note Vpwr sweep range: 0V 12V , and circuit current and voltage values in the table.

Fourth screen: Plots of DC Transfer Curves Tables (.Plot command in file)

Note that each plot is a comparison with sweep input, Vpwr, 0V 12V.
The first graph plots V(2), (0V-9V), vs Vpwr.
The second graph plots V(2,3), (0V-4V), vs Vpwr.
The third graph plots I(R1), (0A-1A), vs Vpwr.

Note: The Probe program will present detailed graphs of these sweep values.

Job Concluded - Total Job Time is the last Browse screen.

Press Esc to exit Browse Output.

II. The Probe Graphical Waveform Analyzer from DOS

Follow these steps to exit the PSpice Control Shell to the DOS prompt.
Select Quit in the Control Shell.
Select "Exit to DOS," and Enter.
Accept "Save Changes..." by pressing Enter or S.
Press Enter as needed to complete the exit.

Probe

At the PSEVAL52 directory, type PROBE DCSWEEP, and Enter. This command calls the Probe Graphical Waveform Analyzer and filename created earlier. Any filename could be called.

The initial Probe menu screen is displayed. We will be using this menu later in operating Probe.

Select Exit, and Enter.

The Probe title screen is displayed. Notice that only one analysis, Dc_sweep, which was specified in the circuit file, is available. The circuit title line is listed along with the temperature setting at the top of the screen.

```
                         | Probe |
               Waveform  Analyzer  for PSpice
                Version  5.2 - July  1992
       (C) Copyright  1985 - 1992 by MicroSim  Corporation

Circuit:  *DCSWEEP...AN  INTRO  TO .DC SWEEP
Date/Time  run:                                  Temperature:  30.0




 Exit_program    Dc_sweep
```

Select Dc_sweep, and Enter to return to the menu screen. Vpwr is on the X_axis. Circuit variables will be plotted using the Y_axis.

```

 (Initial  Probe  menu  screen)




0V                                                        12V
                           Vpwr
 Exit  Add_trace   X_axis   Y_axis   Plot_control   Display_control   Macros
 Hard_copy   Zoom   Label   Config_colors
```

Displaying Sweep Voltages

Select Add_trace, and Enter. Press F4 for a list of variables and expressions.
Notice that all node voltages and resistor currents are available, along with Vpwr and I(Vpwr).

Select Vpwr, and Enter twice.

The screen displays the sweep of the input voltage, Vpwr, 0V to 12V. Next, display the voltage at Node 1.

Select Add_trace, and Enter. Press F4 for variables and expressions.
Select V(1), and Enter twice.

The V(1) trace is superimposed on the Vpwr trace. Both represent the same point in the circuit, Node 1.

Repeat the steps to display V(2) and V(3).
V(2) is the first division of the input voltage and sweeps from 0V to 9V.
V(3) is the second division of the input voltage and sweeps from 0V to 5V.

Select Remove_trace, and Enter.
Select All, Enter, Exit, and Enter.
All traces are removed.

Displaying Sweep Currents

Plot I(Vpwr)

Select Add_trace, and Enter. Press F4.
Select I(Vpwr), and Enter twice.

The current through the source is displayed as 0A to -1A, observing the polarity of the source.

Plot I(R1)

Select Add_trace, and Enter. Press F4.
Select I(R1), and Enter twice.
The current through R1 and throughout the series circuit is displayed as sweeping from 0A to +1A.

Select Remove_trace, All, Exit, and Enter at each selection. All traces are removed.

The following is an introduction to Probe screen menus. The DCSWEEP circuit is used to provide data for introducing the various menu functions.

III. Probe Graphical Waveform Analyzer Menus

The initial Probe Menu screen should be displayed.
Select Add_trace and display Vpwr and V(2) sweep voltages.

X_axis

Select X_axis, and Enter, and note options.

X_axis Options

Exit Log Set_range Restrict_data X_variable Fourier Performance_analysis

Set_range

Select Set_range, and Enter.
At Enter a range: type .1V 12V , and Enter.
Notice the new X_axis range.

Log

Selecting Log and pressing Enter will provide a Log sweep of the X_axis if we have selected a range starting at a value other than 0. Interesting, but of little significance in this linear circuit.
Select Linear, and Enter to return to a linear X_axis plot and to return to the X_axis menu screen.

Set_range

Any range can be entered for observation. For example, if the lower part of the sweep is of major interest, then the X_axis range adjustment can focus the graph to that region.

Select Set_range, and Enter. Set_range: type 0V 6V , and Enter.
The X_axis is drawn to a maximum of 6 Volts. Y_axis is automatically scaled to the new X_axis range.
Select Auto_range, and Enter to return to the full default range.

Restrict_data

Select Restrict_data, and Enter. At Enter a range: type 0V 6V, and Enter. The display is restricted to the limits specified on both scales.

Note: Restrict_data is available only for the input sweep variable, Vpwr.

Unrestrict_data

Select Unrestrict_data, and Enter to reverse the previous adjustment.

X_variable

The selection of X_variable allows setting any circuit variable or expression as the X_axis values.

Select X_variable, and Enter:
Press F4 to display the circuit variables to plot on the X_axis.
Select V(2), and Enter twice.
The graph plots the Y_axis in reference to V(2), the selected X_axis variable.

Reset X_variable to Vpwr:
Select X_variable, and Enter.
Press F4 and select Vpwr, and Enter twice.

Fourier_analysis (N/A this exercise)

The .FOUR control statement is not available. Fourier analysis is used with transient analysis and performs a harmonic decomposition of a specified voltage or current. Outputs of Fourier analysis include the DC component, the fundamental and second to ninth harmonic components of the fundamental of the voltage or current named in the .FOUR statement, i.e., .FOUR 1000 V(7).
See EOC Problem 1 in Chapter 1 for an example of Fourier analysis.

Performance_analysis (N/A this exercise)

Exit the X_axis screen.

Vpwr and V(2) traces should be displayed on the screen from the previous exercise.

We have completed the X_axis menu. Let's take a look at Y_axis options.
Select Y_axis, and Enter.

Y_axis Options

Exit Log Auto_range Set_range Add_axis Change_title Color_Option

Set_range

Select Set_range, and Enter.
At Enter a range: type .1V 12V , and Enter.
The Y_axis range is set to the new values.

Log

Selecting Log and pressing Enter will provide a Log sweep of the Y_axis if we have selected a range starting at a value other than 0. Interesting again, and again of little significance in this linear circuit.
Select Linear, and Enter to return to the Y_axis menu screen.

Set_range

Select Set_range, and Enter. At Enter a range: type 0V 20V
Press Enter.
The Y_axis range is set to the new values.

Y_axis Title

Select Change_title, and Enter.
At Change title: type Volts, and Enter.
The Y_axis is titled as defined.

Select Exit, and Enter to return to the Probe main menu screen.
Select Exit, and Enter, to leave Probe main menu.
At the Probe title screen, select DC_sweep and return to the main menu.

Add_axis

This selection will allow an additional Y-axis graph to display two plot variables simultaneously, i.e., current and voltage.
Select Add_trace, F4, select I(R1) and Enter twice. Vpwr is displayed on the X_axis and I(R1), with a range of OA-1A, is displayed on the Y_axis.
Select Y_axis, and Enter.
Select Change_title, Enter, and type: CURRENT R1 , and Enter.
The Y_axis is titled.
Select Add_axis, and Enter.

The original Y_axis scale moves left and a second scale is ready to be edited. Select Select_axis, and Enter, and use the left/right Arrow keys to select the new axis. The two >> symbols mark the selected axis.

Press Enter.

Select Set_range, Enter, type 0V 12V, and Enter. Exit to the initial Probe screen. Select Add_trace, F4, select V(2), and Enter twice.

The graph is completed and displays both current and voltage on the Y_axis. The legend notes the Y_axis scale to read for each plot.

Select Y_axis, and Enter. Select Change_title and type VOLTS, and Enter. The second Y_axis scale is labeled.

Select Remove_axis, Enter, Exit, and Enter.

Select Remove_trace, Enter, All, Enter, Exit and Enter.

Select Add_trace, F4, select V(2), and Enter twice.

Change_title

Select Y_axis, and Enter.

Select Change_title, and Enter. At Change title: backspace to erase CURRENT R1, and type VOLTS, and Enter.

The Y_axis is titled for voltage display.

Color_Option

Select Color_Option, and Enter, and observe available selections.

Exit   Normal   Match_axis   Sequential_per_axis

Select Match_axis, and Enter.

The VOLTS label color matches the Node 2 trace in the display for easy recognition. Select Color_option, Enter, Sequential_per_axis, and Enter.

Auto_range

Set Y_axis Range: 0V 20V, and Enter. Note the expanded scale.

Select Auto_range, and Enter.

The Y_axis range is changed to 0V to 10V, to better display V(2).

Select Exit, and Enter.

Select Add_trace, and display V(1).

The Y_axis range is automatically changed to 0V to 12V to display the V(1) maximum of 12 Volts.

Select Y_axis, and Enter.

Select Color_option, and Enter.

Select Match_axis, and Enter.

The same color is used for the Y_axis label and both traces.

Each trace is individually marked to match legend.

Select Color_options, Enter, Normal, Enter, and observe screen colors.

The original trace colors are restored.

Sequential_per_axis

Select Color_option, Enter, Sequential_per_axis, and Enter.

Selecting Sequential_per_axis provides a different color for each trace. "Sequential" means the color sequence that was selected in the Config_colors selection of the Probe menu screen.

Select Exit, and Enter to return to the Probe menu screen.

Config_colors

Select Config_colors, and Enter. The screen displays the six default trace colors. The first trace is at the bottom. Selecting Set_color provides a menu for Background, Foreground, Trace, Cursor, and Mouse selection. Press Enter. Select Cursor, and Enter. A Cursor color set screen is displayed. Color is mixed using the Red, Green and Blue color bars. (Note vertical and horizontal cursors) Click on a color and drag left or right to define amount of color mix or click on the color bar at the amount required. Click on Exit to use the new color. Click on Default, then Exit, to accept default settings.

Repeat for Background, Foreground, and Mouse.

Select Trace, and Enter, and follow prompts to edit trace colors. Click on Exit to use new colors. Click on Default, then Exit, to accept default settings.

Select Exit, and Enter to return to the Config_colors screen.

More traces can be added, one at time, by selecting More_trace_colors and pressing Enter. Selecting Fewer_trace_colors and pressing Enter will remove traces at the same rate. Selecting Save saves trace colors to disk. Default will return to the original six colors.

Select Default, and Enter. Exit the Config_color screen.

Note: V(2) and V(1) traces should be displayed from the previous exercise.

Plot_control

Select Plot_control, and Enter.

Plot_control Options

Exit Add_plot Always_use_symbols Never_use_symbols Mark_data_points

Add_plot

Add_plot displays an additional plot above the original graph.

Select Add_plot, and Enter. Select Select_plot, and Enter.

Use the Up/Down Arrow keys to select the upper plot. The >> symbols indicate the selected plot. Accept Exit, and Enter.

Select Exit again and return to the initial Probe menu screen to provide data for this new plot.

Select Y_axis, and Enter. Select Set_range. Set to 0V 10V , and Enter. Select Exit, and Enter.

Select Y_axis, and Enter.
Select Change_title, Enter, type NODE 3, and Enter.
Select Exit, and Enter.

Select Add_trace, F4, V(3), and Enter as needed.
The additional plot displays the voltage sweep at Node 3.
Select Y_axis, Enter, Auto_range, and Enter, for a full-scale display of V(3) on the Y_axis (0V-5V).
Select Exit, and Enter.

Select Plot_control at the initial Probe menu, and Enter.
Select Remove_plot, and Enter, to remove the V(3) plot.
Accept Exit, and Enter.
Select Plot_control, and Enter.

Always_use_symbols

Select Always_use_symbols to place trace symbols that identify particular data readily. Trace colors and symbols assist in reading graphs.

Never_use_symbols

Select Never_use_symbols to remove trace symbols from plots.

Mark_data_points

Select Mark_data_points, and Enter.
Each data point is marked on the trace as defined in the command statement: .DC Vpwr 0 12 1 (Data taken in 1 Volt steps).
Select Do_not_mark_data_points and remove these marks.
Exit Plot_control.

Display_control

Select Display_control, and Enter.

Display Options

Exit Restore Save List_displays View_display_detail One_delete All_delete

Options are available to edit present or previously saved displays.
Select Exit, and Enter to return to the initial Probe screen.

Macros

Macros in PSpice Probe are used the same as other program macros...to define a group of sequentially repeated commands. In Probe, for example, macros are used for grouped waveform calculations. Defining a new macro must include:

<name>[(arg[,arg]*)] = <definition>

Press Esc to exit Macros.

Hard_copy (Not available)

Select Hard_copy, page length, and Enter for print of present display. Hard copy quality and page size in the evaluation package will not be what you request. Press Exc to exit Hard_copy.

Note: V(2) and V(1) traces should be displayed from the previous exercise.

Cursor

Select Cursor, and Enter.

Using the Mouse or Left/Right Arrow keys, specific points on plots can be accurately measured. Select a point on the trace and click and hold the left mouse button and move up and down the plot. Notice the readout at the bottom right of the screen. The first voltage reading is X_axis, the second is Y_axis. Click the left mouse button at any point on the trace for voltage readings at that point.

Select Exit, and Enter.

Zoom

Zoom is useful when selecting specific areas of the screen to view or Hard_copy.

Select Zoom, and Enter.

Select Specify_region, and Enter.

Click the left Mouse button on the upper left starting point of the graph and drag to the lower right stopping point, and release. The screen display will adjust to the new dimensions.

Click on Auto_range to return to the original graph. Start over at the initial Probe menu screen if you run into trouble.

Try the other Zoom and Pan commands and follow prompts. They are interesting and require practice.

Click on Exit to leave Zoom.

Label

The Label selection allows labeling the graph. We will place V(2) and Vpwr labels and a graph title in this exercise.

Exit to the Probe title screen and select DC_sweep, and Enter.

Place Vpwr and V(2) traces on the graph by using the Add_trace option and procedure.

Select Label, and Enter.

Select Text, and Enter, and type a label name to be placed on the V(2) trace. For example, type NODE 2 VOLTAGE, and Enter.

Using the Mouse, click and hold the left button on the label and drag to the right of the V(2) trace. Release the button.

Repeat for Vpwr. Place the label to the left of the trace.

Repeat the above procedure and place the following title lines at the top center of the graph:

PROBE GRAPHICAL WAVEFORM ANALYZER
DC SWEEP ANALYSIS

Select Exit, and Enter.

Select Exit, and Enter, to leave Probe menu screen.

Select Exit_program, and Enter, to leave the Probe Graphical Waveform Analyzer.

The DOS prompt should be displayed.

Conclusion of Chapter 3.

EOC Exercises

(1) Repeat this chapter to gain experience with Probe and the .DC sweep control statement.

Chapter 4

STIMULUS EDITOR

Objective: (1) To introduce the Stimulus Editor
(2) To survey StmEd variables
(3) To enter StmEd from the Control Shell and modify the circuit file

Notes: A short wait in some parts of the program may be necessary if a coprocessor is not installed. Avoid pressing the Enter key more than once during these delays. A second Enter will be executed next and you may move to an unfamiliar screen. Double clicking with the Mouse should also be avoided for the same reason.

"Select" means to use the Arrow keys or Mouse to highlight a choice on the screen.

I. Using the Stimulus Editor from the DOS prompt

The Stimulus Editor, or StmEd, is the input "generator" to the circuit under test. It is an independent program accessible directly from DOS or operated within the PSpice Control Shell. First, we will enter the Stimulus Editor from DOS and create a new circuit file named BJTAMP. Then, we will edit the inputs or "stimuli" to this amplifier. We can set up the inputs even though the circuit does not exist. It is like adjusting a generator to use later when the circuit is ready to be tested. Follow these steps to place StmEd into operation from DOS.

At the PSEVAL52 directory, type STMED, and Enter.

An existing file can be named to be modified or a new file name may be entered. Our file is a new one.

At "Enter circuit file name:" type BJTAMP, and Enter.

At "The file BJTAMP.CIR does not exist: Create (Y/N)?", type Y, and Enter, to create the BJTAMP.CIR file. We are now at the initial StmEd menu screen.

Note: If BJTAMP.CIR already exist, exit to the PSEVAL52 prompt and type ERASE BJTAMP.CIR, and Enter.

If we had named an existing circuit file, StmEd would have read the stimuli (the inputs and waveforms) and displayed them at this first screen. We could have edited these stimuli and saved the modified file.

At this point we have created a new file called BJTAMP. Next, we will edit an input to be used with this file.

The screen displays a graph with time (0s-1.0us) as the X_axis. StmEd is programmed to accept a time varying stimulus to the circuit. This initial screen also provides a list of options.

Initial StmEd Options:

Exit **New_stimulus** **Plot_control** **X_axis** **Y_axis** **Hard_copy**

Defining the Input Stimulus

Select New_stimulus, and Enter.

The input to the new circuit can now be defined.
At "Enter the stimulus device name (U, I or V):", type Vin , and Enter.

The input signal is named Vin.

Press Enter to accept "SIN," for sine wave input as the Transient spec type.

The Stimulus Editor is now ready to set the VOFF, the offset voltage of the input stimulus.
A value must be entered when editing...no blank spaces. Enter values exactly as shown.
At VOFF: type .25V, and Enter.
VOFF is set to .25V.

As each parameter is displayed, enter the following data:

VAMPL (Voltage input amplitude)	: .5V
FREQ (Frequency)	: 1000 (Omit Hz)
TD (Time delay)	: 0
DF (Damping factor)	: 0
Phase	: 0

At the last entry, Phase, the StmEd screen displays a 1KHz sine wave, with Vpeak= .5V, and VOFF= .25V.

Observe the waveform. Study the Specs table.
(Sketch the waveform on page 4 - 8, Problem 3)

Vin parameters will be adjusted in the next steps.
The full Modify_stimulus menu is displayed. Study the options available.

Modify_stimulus Options:

Exit Transient_parameters Spec_type Other_info X_axis Y_axis Display_help Hard_copy Cursor

If we exit to the StmEd main menu or "home" screen, the following options will be displayed.

StmEd Main Menu Options:

Exit New_stimulus Modify_stimulus Delete_stimulus Plot_control Y_axis Hard_copy Cursor

Selecting New_stimulus will allow the user to name and edit another input to the circuit. Modify_stimulus presents the previously entered input for editing. Select Modify_stimulus, and Enter, to return to the Vin display.

Modifying Vin, the New Stimulus

In the next part of the exercise we will re-edit the input signal, Vin, to different specifications. It will be necessary to exit the Transient_parameters menu to complete the editing.
First, set the offset voltage, VOFF, to 0 Volts. Enter all values as shown.

(A) Select Transient_parameters, and Enter.
Select VOFF, and Enter.
Backspace to erase, type 0, and Enter.
Select Exit to leave the Transient_parameters screen, and Enter.

The StmEd screen displays the original input voltage and waveform with the modified VOFF= 0V. Next, edit the amplitude of Vin.

(B) Select Transient_parameters, and Enter.
Select VAMPL, and Enter.
Backspace to erase the original .5V and type 1V, and Enter.
Exit the Transient_parameters screen.
StmEd displays the new Vin= 1Vpeak.

(C) Select Transient_parameters and change the frequency
(FREQ) of the input to 1500 Hz.

Enter after editing.
Exit the Transient_parameters menu screen.

Three hertzs of the new input frequency are graphed on the StmEd screen. Using the time baseline and f= 1/t, calculate and verify the frequency.
Next, delay the start of the input waveform by editing TD, time delay.

(D) Select Transient_parameters and set TD (time delay) to
.4ms (milliseconds).
Enter after editing.
Exit the Transient_parameters menu screen.
StmEd displays Vin, delayed by .4ms.
Next, set the damping factor, DF.

(E) Select Transient_parameters and set DF (damping factor)
to 1000.
Enter after editing.
Exit the Transient_parameter menu screen.
StmEd displays the 1500Hz waveform with the prescribed damping.
Next, set the signal phase to 45 degrees.

(F) Select Transient_parameters and set the phase at 45 degrees.
Enter after editing.
Exit the Transient_parameter menu screen.
The new stimulus is shifted 45 degrees.

Remove all Transient_parameters except edit VAMPL= .5V and leave FREQ= 1500. Select Transient_parameters, press Enter, then select the parameter to be modified. When editing is completed, select Exit to leave the Transient_parameter menu screen and return to the modified display. The screen should display the 1500Hz waveform with Vp= .5V.

More Modify_stimulus Options

Select "Spec_type" input to specify the type of input. Sine wave is the only stimulus option available.

Press Enter to accept sine wave input.

Select Display_Help, and Enter.
Review these Help screens. Notice that the specs for input signal, Vin, are listed on each page.
What does an asterisk before a parameter name indicate?
Select Exit when finished, and Enter.

Note: Hard_copy is not available. If Hard_copy is selected the program may lock up. If this happens you may have to begin again at the PSEVAL52 directory.

Select Cursor, and Enter.
Exact measurements of the waveform are possible by using the Mouse or Left/Right Arrow keys. Move the cursor crosspoint around and notice the values in the table at the lower right of the screen. C1, first reading is X_axis, in microseconds and milliseconds; second reading is Y_axis in millivolts.
Again, Hard_copy is not available.
Select Exit, and Enter to return to Modify_stimulus menu screen.

Select Y_axis, and Enter. Select Set_range, and Enter.
At "Enter a range:" type -2V 2V, and Enter.
Exit and observe the new vertical scale.
Select X_axis, and Enter. Select Set_range, and Enter. At "Enter a range:" type 0s .67ms, and Enter. StmEd presents a new display of one hertz of the 1.5 KHz sine wave input.
Select Auto_range, and Enter. The program's Auto_range time base should be presented.
Select Exit, and Enter.
Select Exit, and Enter, to leave the Modify_stimulus screen.

New_Stimulus Options

Selecting New_stimulus allows editing of as many inputs as necessary to analyze circuit behavior.
Select New_stimulus, and Enter.
At "Enter the stimulus device name:", type V2 , and Enter.
Accept sine wave input by pressing Enter at "SIN."
Set: VAMPL= .25V, FREQ= 2500, Phase= 45. Press Enter.
The V2 input and spec table are displayed.
Select Exit, and Enter, to leave New_stimulus menu screen.
StmEd displays the two input signals with the various parameters that were specified. (Sketch the two waveforms on page 4 - 9, Problem 4)

Enter a third New_stimulus, V3, by editing VOFF= 0V, VAMPL= .4V, FREQ= 1200, and Phase= 60 degrees. Press Enter at each parameter setting.

Observe the new V3 stimulus.

Select Exit, and Enter, to leave New_stimulus menu screen. Study the waveforms.

In StmEd, one input or several inputs can be displayed simultaneously and then modified separately. The simultaneous display of inputs allows on-screen modifications to coordinate time dependent inputs. Stated another way, if the circuit under test requires two or more inputs with specific related variations, StmEd has the capability of editing these variations. In some test situations, this would not be possible with other stimuli sources.

Select Modify_stimulus and Enter.

At this screen, an individual stimulus can be selected for editing. Use the Arrow keys to highlight one of the inputs, then press the spacebar to select it. Press Enter.

StmEd presents an editing screen for the selected input.

Leave the parameters as set and exit.

Select Exit, and Enter, to go to the Stimulus Editor title screen. Here, we have three options.

Selecting Start_Editor will re-enter the Stimulus Editor.

Selecting Abort_program will ask if you wish to save file (Y/N)?

We will choose the third option, Exit_program.

Select Exit_program, and Enter, and the new BJTAMP.CIR file will be saved to disk. Now, let's return to StmEd.

At the PSEVAL52 directory, type StmEd, and Enter.

At Enter circuit filename: type BJTAMP , and Enter. The main StmEd menu is displayed with the three stimuli that were set for the BJTAMP circuit.

Select Exit, and Enter.

The next exercise will demonstrate stimulus editing and saving the BJTAMP.CIR file using the Stimulus Editor from the PSpice Control Shell.

Select Exit_program, and Enter.

II. Using StmEd from the Control Shell to Modify_Stimulus

We will now enter StmEd from the Control Shell and modify the Vin input in the BJTAMP file. These changes will be saved to the circuit file.

Type PS at the PSEVAL52 directory, and Enter. The PSpice Control Shell menu screen is displayed.

Select Files, and Enter.

At Current file, type BJTAMP, and Enter.

First, we will take a look at the three inputs we previously edited into the

BJTAMP circuit file. Remember, we have not defined the circuit, just the input stimuli...Vin, V2, and V3.

At the PSpice Control Shell, select Files, Enter, Edit, and Enter. The inputs are defined. Study each line and take note of the parameter sequence used in editing. Can you name the sequence...VOFF, VAMPL, FREQ, TD, DF, and PHASE?

PNode and NNode are node indicators and will be replaced with node labels (i.e., 1, 2, 3, IN, etc.) when the inputs are edited into the circuit file. PNode is the positive terminal, NNode the negative terminal.

Take a close look at the Vin input line. Adjustments will be made to this input. Press "Esc" and "D" to exit.

Select StmEd, Edit, and Enter.

Accept changes if requested and press Enter.

The display is an "O'scope type" graph with Vin and the two other inputs...V2 and V3.

Select Modify_Stimulus, and Enter.

Arrow to Vin and press the spacebar to select this input. Press Enter to view and adjust Vin. The screen displays the Vin waveform only, with a Transient Spec table ready for modification.

Select Transient_parameters, and Enter.

Set each parameter as follows:

VOFF(offset voltage)	= .25V
VAMPL(peak amplitude)	= 1.25V
FREQ (frequency)	= 1500
TD(Time delay)	= .4ms
DF(Damping factor)	= 200
Phase	= 60

Notice how each entry is registered in the Transient Spec table.

Select Exit, and Enter.

Study the new input stimulus. (Sketch the modified Vin waveform on page 4 - 9, Problem 5)

Select Exit to leave the Modify_parameters menu screen.

The new Vin settings are incorporated into the display. The three stimuli, including the modified Vin input will be saved to the BJTAMP circuit file when we exit StmEd.

Select Exit, and Enter, to leave StmEd main menu.
Select Exit_Program to leave StmEd, and Enter. The PSpice Control Shell is displayed.
Select Files, Enter, Edit, and Enter.

The BJTAMP.CIR file is displayed with the modified Vin input.

```
Vin  PNode  NNode
+SIN( .25  1.250  1.500E3  400.0E-6  200  60)
```

Locate offset voltage, amplitude, frequency, time delay, damping & phase in the statement. Study each of the circuit file stimuli statements.
Press "Esc," and "D", to exit to the Control Shell.
Select Quit, and Exit to DOS, and Enter.

Note: Erase the circuit file used in this chapter by typing ERASE BJTAMP.CIR at the PSEVAL52 prompt.

Conclusion of Chapter 4

EOC Problems

1. Define, in a step-by-step process, the two ways described in this chapter to access StmEd.

2. Repetition builds retention. Repeat this chapter to become even more familiar with StmEd.

3. Sketch the Vin waveform from page 4 - 2. Label the graph.

4. Sketch the waveforms from page 4 - 5. Label the graph.

5. Sketch the modified Vin waveform from page 4 - 7. Label the graph.

Notes:

Chapter 5

THE DEVICE LIBRARY

PARTS PARAMETER ESTIMATOR

Objectives: (1) To introduce the PSpice Device Library
(2) To introduce the PSpice Parts Parameter Estimator

Notes: A short wait in some parts of the program may be necessary if a coprocessor is not installed. This is especially true in the PARTS program. Avoid pressing the Enter key more than once during these delays. A second Enter will be executed next and you may move to an unfamiliar screen. Double clicking with the Mouse should also be avoided for the same reason.

"Select" means to use the Arrow keys or Mouse to highlight a selection in the menu.

If the PARTS program stalls, see note at end of chapter.

I. PSpice Device Library

The control statements to reach device models (.Model) and subcircuits (.SUBCKT) in the following PSpice libraries are:

.LIB	- searches the master library files
.LIB EVAL.LIB	- searches the evaluation library files

ASCII text files in these libraries are easily entered and edited.

The PARTS program is used to create another library called "User Library" to include specialized models and subcircuits. (See Chapter 1, Figure 1)

The Model Library in the production version of PSpice contains over 4,600 analog devices and more than 1,600 digital devices. Examples include:

Diodes	Triacs
Bipolar Junction Transistors	Voltage regulators
Power MosFET's	Zener Diodes
Small-signal JFET's	TTL Devices
Operational Amplifiers	High-Speed CMOS
Transformer cores	ECL Devices
Optocouplers	PAL Devices
Voltage comparators	Analog/Digital Interface

An overview of the reduced version of the library in the evaluation package (EVAL.LIB) will help us comprehend the massive amount of programmed data necessary to simulate various devices and subcircuits. The following is a summary of device models and subcircuits contained in the evaluation library.

Part Name	Part Type
Q2N2222A	NPN bipolar transistor
Q2N2907A	PNP bipolar transistor
Q2N3904	NPN bipolar transistor
Q2N3906	PNP bipolar transistor
D1N750	Zener diode
MV2201	Voltage variable capacitance diode
D1N4148	Switching diode
MBD101	Switching diode
J2N3819	N-channel JFET
J2N4393	N-channel JFET
LM324	Linear Operational Amplifier
UA741	Linear Operational Amplifier
LM111	Voltage comparator
K3019PL_3C8	ferroxcube pot magnetic core
KRM8PL_3C8	ferroxcube pot magnetic core
K502T300_3C8	ferroxcube pot magnetic core
IRF150	N-type power MOSFET
IRF9140	P-type power MOSFET
7402	TTL digital 2-input NOR gate
7404	TTL digital inverter
7405	TTL digital inverter, open collector
7414	TTL digital inverter, schmidt trigger
7474	TTL digital D-type flip-flop
74107	TTL digital JK-type flip-flop
74393	TTL digital 4-bit binary counter
A4N25	Optocoupler
2N1595	Silicon Controlled Rectifier
2N5444	Triac

This sample library contains approximately 25 pages of device model data and comments. Keep in mind that device model parameters were taken from manufacturer's data sheets. As an example, the following parameters characterize the 2N2222 bipolar junction transistor.

```
.Model Q2N2222A NPN(Is=14.34f Xti=3 Eg=1.11 Vaf=74.03
+                   Bf=255.9  Ne=1.307  Ise=14.34f  Ikf=.2847
+                   Xtb=1.5 Br=6.092  Nc=2 Isc=0  Ikr=0  Rc=1
+                   Cjc=7.306p  Mjc=.3416  Vjc=.75  Fc=.5
+                   Cje=22.01p  Mje=.377  Vje=.75  Tr=46.91n
+                   Tf=411.1p  Itf=.6  Vtf=1.7  Xtf=3  Rb=10)
*                   National        pid=19        case=TO18
*                   88-09-07 bam    creation
```

The library .model defines a typical device. Variations from typical can be listed in the circuit file (netlist) .model statement. Default values will be used if specific parameters are not listed in the .model statement.

BJT parameter definitions are included in Appendix B.

Library Linear IC Definitions

A typical linear IC library listing is the UA741 operational amplifier subcircuit:

```
* connections:   non-inverting  input
*                | inverting  input
*                | | positive  power  supply
*                | | |  negative  power  supply
*                | | | | output
*                | | | | |
.subckt UA741  1 2 3 4 5
*
c1    11  12 8.661E-12
c2     6   7 30.00E-12
dc     5  53 dx
de    54   5 dx
dlp   90  91 dx
dln   92  90 dx
dp     4   3 dx
egnd  99   0 poly(2)  (3,0) (4,0) 0 .5 .5
fb     7  99 poly(5)  vb vc ve vlp vln 0 10.61E6  -10E6
+            10E6 10E6  -10E6
ga     6   0 11 12 188.5E-6
gcm    0   6 10 99 5.961E-9
iee   10   4 dc 15.16E-6
```

Cont:

Cont: UA741 op amp subcircuit library listing:

```
hlim  90   0 vlim  1K
q1    11   2 13 qx
q2    12   1 14 qx
r2    6   9 100.0E3
rc1    3  11 5.305E3
rc2    3  12 5.305E3
re1   13  10 1.836E3
re2   14  10 1.836E3
ree   10  99 13.19E6
ro1    8   5 50
ro2    7  99 100
rp     3   4 18.16E3
vb     9   0 dc 0
vc     3  53 dc 1
ve    54   4 dc 1
vlim   7   8 dc 0
vlp   91   0 dc 40
vln    0  92 dc 40
.model dx D(Is=800.0E-18  Rs=1)
.model qx NPN(Is=800.0E-18  Bf=93.75)
.ends
```

Op Amps are modeled at ambient temperature, 27 degrees C, and will not track changes in temperature. The evaluation library contains models for nominal, not worst case, devices.

Notice the UA741 subcircuit listing begins with the required .SUBCKT statement and closes with the .ENDS statement.

Viewing and Printing the Evaluation Library

The evaluation library file may be viewed on the screen by entering the following command at the PSEVAL52 directory:

"TYPE EVAL.LIB"

Press the PAUSE key to stop scrolling, any key to continue scrolling.

To print a copy of this library, type,

"PRINT EVAL.LIB", and Enter.

We will take a tour of the PARTS program in the following section of this chapter.

II. PARTS PARAMETER ESTIMATOR

The next version of the Model Library (6.0) will include over 9,000 devices with the addition of above 1700 Japanese devices, and other device models can be added as needed by the user. A full version of PSpice will, no doubt, have all devices on file that we would need in the education environment but would not be an inexhaustible source that would meet the design requirements of every-one. The PARTS program helps fill the gap between existing library models and specialized devices and is limited only by the user's ability to define the model required.

PARTS.EXE is an independent program in PSpice and is operated as a stand alone package from the DOS prompt. (See Chapter 1, Figures 1 & 2) This progam enables the user to create new device .model definitions by gathering infor-mation from manufacturer's data sheets. Models can, of course, be written directly into the circuit file (netlist) but it is much more convenient to use PARTS and place them into library files to be called when needed. Subcircuits (.Subckt to .Ends statements) can also be written, filed, and used. New parameter values are estimated and verified in PARTS, including best and worst case models with device and temperature variations. The user-friendly interactive screens allow device parameter adjustments and display analysis results. The following condensed list of devices and screens that demonstrate device characteristics illustrate the detailed analyses available in the PARTS program in PSpice.

DEVICES	SCREENS PROVIDED
Diodes	Forward DC voltage vs current Reverse leakage current Breakdown voltage Reverse recovery time
BJTs	Gain vs collector current Small signal output admittance Ic vs Cce(sat)
Small-signal JFETs	Transconductance vs Id Miller capacitance DC Id vs Vgs
Power Mosfets	Small signal transconductance vs drain current DC Id vs Vgs Switching time Turn-on charge Output capacitance

Cont:

Cont:

DEVICES	SCREENS PROVIDED
Op Amps	Large signal output swing and slew rate Open loop gain
Voltage Comparators	Open loop gain, input bias current Falling delay Transition time

Operating the Parts Parameter Estimator program

At the PSEVAL52 directory,

Type PARTS, and Enter.

The screen displays the initial Parts Parameter Estimator menu. Here, two choices are available. One, a device can be completely modeled and saved to a library file by the user. And two, a model can be called from the library and modified to user design specifications.

The diode selection is the only option available in the evaluation version.

```
                         |PARTS |
            Parameter estimator for PSpice models
                  Version 5.2 - July 1992
          (C) Copyright 1991 by MicroSim Corporation
          ________________________________________

0) Exit Program
1) Diode     (signal/rectifier/Zener)          <--only model available
2) Bipolar Transistor (gen'l purpose)             in Evaluation
3) JFET (small-signal, gen'l purpose)
4) Power MOSFET Transistor (all types)
5) Operational Amplifier (bipolar/FET)
6) Voltage Comparator    (bipolar OC)
7) Nonlin. Magnetic Core (ferrite/MPP)

Select: 1
```

Press Enter to accept 1) Diode modeling.
At Device part number (or name): type 1N4001, and Enter.

Note: The parameters listed and screens provided are not necessarily 1N4001 quantities but are used for demonstration purposes only.

Diode parameters will be displayed on PARTS interactive screens. The screens define the diode in data sheet quantities and model definitions. Study each display in terms of X_axis and Y_axis values and ranges.

Notice the menu list at the bottom of this first screen, and the device name (D1N4001) and specific screen name in the upper right corner. The small block on the right displays model parameters. These parameters can be adjusted to meet specifications as models are written for devices.

Record each of the model parameter values. You will need them later if a quantity is inadvertently changed.

Saturation current,	IS=____	Emission coefficient,	N=____
Series resistance,	RS=____	High injection "knee" current,	IKF=____
IS temp coefficient,	XTI=____	Activation energy,	EG=____

The Diode Curve

Device Curve: Vfwd, Ifwd

The first screen is a graph of Forward Current, Ifwd, vs Forward Voltage, Vfwd, at 27 degrees Celsius. The silicon diode barrier potential "knee" is recognized at approximately .7V. Follow these steps to shape the curve to a more familiar looking plot.

Select Y_axis, and Enter.

Select Linear, and Enter. (Omit this step if the Y_axis is already a linear plot. If the plot is linear, linear will not be available in the Y_axis menu options)

Select Set_Range, and Enter.

Type at Range: 0A 3A, and Enter, then exit.

If the X_axis range is not 0V - 2V, follow these next steps.

Select X_axis, and Enter.

Select Set_Range, and Enter.

Type at Range: 0V 2V, and Enter.

The display trace has more of a diode upward swing past the maximum forward diode current.

Screen_info

Select Screen_info, and Enter.

The screen displays diode notations and information. This information identifies the Device Curve (Ifwd vs Vfwd) and model parameters. Study each diode characteristic. Select Exit, and Enter.

Model_parameters

Select Model_parameters, and Enter.

Diode parameters that are available to set to model specifications are:

1)IS 2)N 3)RS 4)IKF 5)XTI 6)EG

Can you identify each abbreviation?

A value set screen is provided when a parameter is selected and Entered. Leave unaltered. Press Esc to exit value set screens.

Select Exit, and Enter.

Hard_copy

Hard_copy is not available in the evaluation version.

Adding Traces

Select Trace, and Enter. Select Add_trace, and Enter.

At Enter value of Temperature: type 50, and Enter.

The PSpice Parts program will add a trace indicating diode operation at 50 degrees Celsius.

Select Trace, and repeat for 10 degrees C.

Study these curves for effects of ambient temperature on diode operation. Barrier potential in silicon diodes is .7V at 27 degrees Celsius. This potential decreases 2mV for each degree rise as more free carriers are created, thus reducing the depletion layer width. The net result is increased diode current with increased temperature.

Select Trace, then Delete_trace, and remove the added traces. Arrow to the trace to be removed and Enter.

Omit Trace_variable in the Trace menu.

Omit Device_curve in the main menu.

Other Interactive Screens

As we move through the following screens, study the device model parameters being estimated and the options in the menu list at the bottom of the screen.

Each of the following screens will require various amounts of time to execute, depending on your computer. If you press the Enter key while the program is running, the extra "Enter" will be processed at the end of the new screen presentation and you may move into an unfamiliar option. If you inadvertently lose your location in the exercise, you may need to exit the PARTS program entirely, then begin again.

(1) Select Next_set, and Enter.

Device curve: Vrev, Cj

A Junction Capacitance, Cj, vs Reverse Voltage, Vrev, graph is displayed. (Cj at 27 degrees C) Record the parameter values.

Model parameters: CJO=______, M=______, VJ=______, FC=______.

Select Model_parameters, and Enter. Model_parameters in each PARTS screen can be selected to set quantities to model specs. Leave unaltered. Press Esc to exit.

Select Screen_info, and study the definitions of device curve names and model parameters. Record these definitions and exit.

Select Trace, Add_trace, and set temperature to 50 degrees C, and study the traces in the display.

(2) Select Next_set, and Enter.

Device curve: Vrev, Irev

A Reverse Leakage Current, Irev, vs Reverse Voltage, Vrev, graph is displayed. Record the model parameter values: ISR=______, NR=______.

Select Screen_info, and study the definitions of device curve names and model parameters. Record these definitions and exit.

Select Trace, Add_trace, and set temperature to 50 degrees C, and study the traces on the screen. What effect does increased temperature have on reverse leakage current?

(3) Select Next_set, and Enter.

Device curve: IBV, BV

Reverse Breakdown is graphed as Reverse Breakdown Current, IBV, vs Breakdown Voltage, BV.

Record the model parameter values: BV=______ , IBV=______.

Select Screen_info, and study the definitions of device curve names and model parameters. Record these definitions and exit.

Select Trace, Add_trace, and set temperature to 50 degrees C, and study the traces in the display.

(4) Select Next_set, and Enter.

Device curve: Reverse Recovery.

A Reverse Recovery graph is displayed with a Forward Current, Ifwd, curve on a Time graph. Study the display.

In this waveform, the applied voltage is switched from forward to reverse bias causing the current to switch from +10mA to -10mA. Recovery time is determined by the amount of "stored charges" in the diode when forward biased (electrons in P-region and holes in N-region). The flat negative portion of the waveform measures the time required for the charges to be removed and is reverse recovery time, Trr.

Record device data: Trr=______, Ifwd=______, Irev=______, Rl=______.
Record the model parameter value: Transit Time, TT=______.
Select Screen_info, and study the definitions of device curve names and model parameters. Select Exit, and Enter.

(5) Select Next_set, and Enter.

A "Model completed" screen is displayed.

NOTE: DO NOT SAVE RESULTS.

Select First_set, and Enter.
PARTS initial screen should be displayed.

From this short tour through the interactive screens displaying diode characteristics and parameter adjustments, we can estimate the tremendous value of the PARTS program when identifying, defining, and adjusting BJT, small-signal JFET, Power Mosfet, Op Amp, and voltage comparator models.

Select Exit, and Enter.

Type 0, and exit the PARTS.EXE program.

Note: Have your instructor erase files D1N4001.MDT and D1N4001.MOD. These files were created by the Parts Parameter Estimator program during these exercises.

Conclusion of Chapter 5

PARTS program note

If the program should stall at the initial PARTS screen, the instructor should erase the unusable D1N4001 files created by the previous user. Specifically, files D1N4001.MDT and D1N4001.MOD.

EOC Problem

(1) Repetition builds retention. Repeat the PARTS Parameter Estimator program exercise to become even more proficient in its operation.

Chapter 6

SERIES CIRCUITS

BRANCH CURRENTS

Independent Current Sources

Objectives: (1) To introduce fundamental PSpice circuit simulation and analysis
(2) To examine Independent Current Sources in PSpice

Electronic component simulation and circuit analysis procedure is the same in small or large experiments. The same basic steps of evaluation are necessary in all testing. Look for these fundamentals in the following simple circuits. And also look for basics in the PSpice computer aided design process.

I. A Series Circuit Introduction

The series circuit in this project requires the following equipment and components:

Power Supply, 0-30 Volts	R1	500 Ohms
Voltmeter	R2	1000 Ohms
Ammeter	R3	1500 Ohms

PSpice will simulate these components and the power supply and test equipment normally used when breadboarding circuits. Next, circuit analysis will be performed in a "PSpice Run." And finally, the program will furnish a printed report of node voltages, source current and power, and any other circuit measurements requested.

The first steps in simulation are to define the objectives, design and sketch the circuit, and appoint node labels to each of the connections in the circuit. Figure 1 is a diagram of the series circuit to be simulated. The voltage source, Vpwr, is connected at Node 1 and Node 0. Node 0 is ground or reference for PSpice circuits.

Note: It may be a good idea to move clockwise with node label numbers in these initial circuits. In larger circuits node numbers will tend to increase from left to right.

The Test Circuit

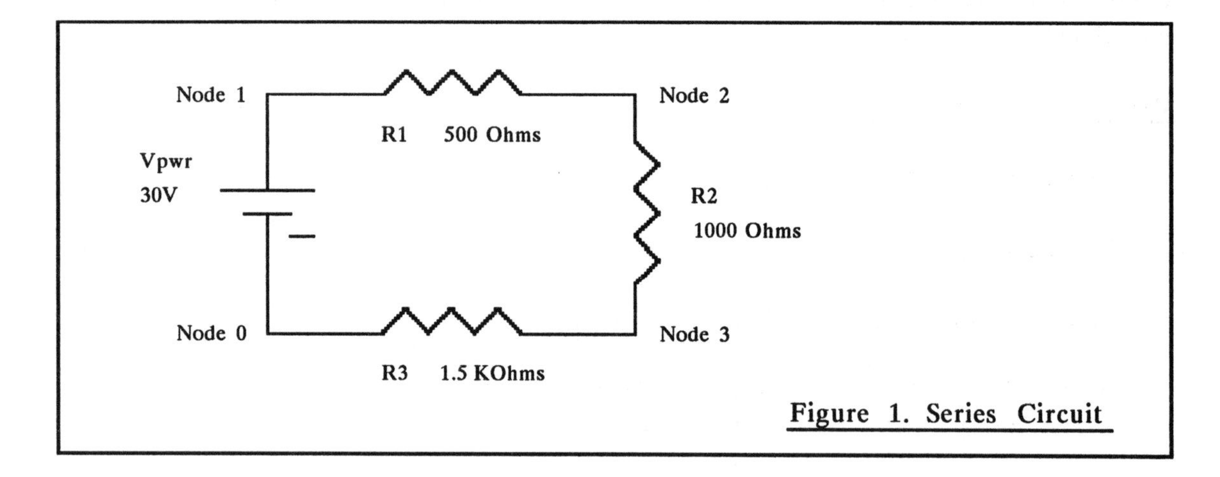

Figure 1. Series Circuit

Circuit Calculations

PSpice does not actually design circuits, but rather simulates and analyses circuit operation for the designer. The designer should complete all calculations to determine circuit expectations. Then, a circuit file list, called a netlist, is used to define the circuit. This netlist is a computer assisted "breadboard" in PSpice. The Circuit Editor in the Control Shell is used to write this netlist and PSpice will interpret the netlist into the actual computer program to simulate and analyze the circuit. The proof of the design is, as always, in the actual operation of the circuit in the test environment. Calculations for the circuit in Figure 1 are as follows:

Vpwr= 30V, Rtotal= 3KOhms
Ptotal= .3Watts
Circuit Current= I(R1)= 10mA

VR1= V(1,2)= 5V
VR2= V(2,3)= 10V
VR3= V(3,0)= 15V

V(1)= 30V
V(2)= 25V
V(3)= 15V

PSpice automatically calculates voltage source current and power dissipation. All node voltages are also reported in the Output file (.OUT) with a default setting. Later, we will sweep the input voltage, Vpwr, and request other voltage and current measurements by writing control statements in the circuit netlist.

PSpice Simulation and Analysis

At the PSEVAL52 prompt, type

PS, and Enter.

The Screen should display the PSpice Control Shell.

Select Files, Enter, Current File, and Enter.
At Circuit file name: type SERIES1, and Enter.

Write the following netlist for this series circuit. Observe node locations. (Line notations and definitions are discussed in Chapter 1)

The Netlist

```
*Series Circuit No.1, Circuit description
Vpwr  1  0  DC  30Volts
R1    1  2  .5KOhm
R2    2  3  1000Ohms
R3    3  0  1.5K
*Control statements
.End
```

Netlist Notes

Different ohms notations are used to illustrate that PSpice is not sensitive to how ohms values are listed. In fact, if the notations were omitted, PSpice would designate any number following a resistor (R) entry as the ohms value. However, it is important not to space between the value listing. For example, 10 KOhms is not acceptable.

All PSpice circuit netlist require an .end statement to signal the last line in the program. The period (.) before .end is also required, as with other control statements or an error message will stop the simulation.

No other control statements are needed in this first "PSpice run." After the completion of the analysis we will take a look at the Output file to observe node voltages, source current and power. Again, these circuit measurements are automatically provided in PSpice.

Press Esc to exit the Circuit Editor and save the circuit file (SERIES1.CIR). Initial PSpice circuit simulation takes place when files are saved, therefore, error messages first appear at this step.

ERRORS!

If you did not have errors, congratulations! Now, let's introduce a few...one at a time. An error is introduced in each of the following netlist changes. Press Esc to save the circuit file after editing in each error. If "Errors" is not indicated, go ahead and attempt to Run PSpice analysis. An error message will be generated. See note below. Notice that most of the titles in Control Shell are dimmed when indicating an error in the netlist.

There are two ways to view errors. One, press F6. And two, arrow to the "Circuit" popdown menu, and press Enter. At Errors, press Enter again, and read the error message. PSpice will list the line where the error is located. Go to the line mentioned and correct the error that was introduced. Repeat for each of the errors listed.

Notes: If an error has been inserted and the program does not signal an error message, then go ahead and attempt to "Run PSpice." Error messages will be printed in the Output file. You will receive a prompt to look at the Output file for these messages.

The first line in the netlist is reserved for the title or name of the circuit. The first numbered line follows the title line. For example, Vpwr is on line #1.

Error list:

1. Remove the (.) from the .End statement.
2. Place a space between .5 and KOhms in line #2.
3. Remove the comment asterisk, (*), from line #5.
4. Omit a node number in either of the circuit description lines.
5. Omit R1 from circuit description line 2.

Correct all errors and proceed.

Run PSpice

Select Analysis, and Enter.
Select Run PSpice, and Enter.

Browse Output

Select Files, and Enter.
Select Browse Output, and Enter, to examine the Output file, SERIES1.OUT.

Use PgDn, PgUp, and the Arrow keys to move about in the Output file. Compare the Output file values with the circuit calculations listed on page 2.

The Output File

Page	Circuit Information
1	The CIRCUIT DESCRIPTION is copied to page 1. (A professional approach to engineering lab reports)
2	SMALL SIGNAL BIAS SOLUTION TEMPERATURE= 27 DEG C V(1)= 30V V(2)= 25V V(3)= 15V (A list of node voltages is presented)
3	VOLTAGE SOURCE CURRENTS Vpwr, CURRENT= -1.000E-02 (Minus indicates source polarity, - to +) TOTAL POWER DISSIPATION 3.00E-01 WATTS JOB CONCLUDED TOTAL JOB TIME

End of Output file

Press Esc to exit Browse Output.

Netlist Editing

Edit the following lines into the SERIES1 netlist. Place the lines between the *Control statement comment line and .End statement.

```
.DC Vpwr 30 30 1
.Print DC V(1,2) V(2,3) V(3,0)
.Print DC I(R1)
```

A .DC sweep statement is necessary to provide the data requested in the .print statements. The first line calls for a DC sweep of Vpwr in 1Volt increments. But since the beginning and end of the sweep are both 30Volts, the sweep will be constant.

Resistor voltages are requested in the first .print statement. For example, V(1,2) is the voltage difference between Node 1 and Node 2 and is the voltage drop across R1. The second print line will provide a measurement of R1 current during the sweep. I(R1) is, of course, the circuit current and could have been included in the first print statement.

Press Esc to save the file.

Run PSpice

Select Analysis, Enter, Run PSpice, and Enter.

Note: The program will beep at you with what appears to be an error message. Everything is A-OK. You are just being informed that there is insufficient data to run Probe. The sweep is constant and Probe needs variables.

Browse Output File

The Output file provides the .print data requested.

Page	Circuit Information			
1	CIRCUIT DESCRIPTION			
2	DC TRANSFER CURVES		TEMPERATURE 27 DEG C	
	Vpwr	V(1,2)	V(2,3)	V(3,0)
	3.000E+01	5.000E+00	1.000E+01	1.5000E+01
3	DC TRANSFER CURVES		TEMPERATURE 27 DEG C	
	Vpwr	I(R1)		
	3.000E+01	1.000E-02		

Press Esc to exit Browse Output.

Netlist Editing

Select Files, Enter, Edit, and Enter.
Edit the .DC sweep line to,

```
.DC Vpwr 0 30 1
```

The sweep begins at 0V and stops at 30V, stepping in 1V increments. Large tables of data will be saved in the Output file. Every value requested in .print statements will be printed at each of the 1V increments. PSpice will also furnish these values in data files (.DAT) for Probe, and Probe will be available after this PSpice run.

Press Esc, and save the edited netlist.

Run PSpice

Select Analysis, Enter, Run PSpice, and Enter.

Probe Graphical Waveform Analyzer

Probe can now be used to observe circuit response to the input sweep. Data requested in .print statements will be graphed.

Voltage Measurements

Probe's initial menu screen should be displayed.
Select Add_trace, and Enter.
Press F4 for a list of variables.
Arrow to Vpwr, and Enter twice.
Vpwr, the input sweep, is graphed.

Select Add_trace, press F4, select V(1), and Enter twice.
Repeat for V(2) and V(3).
Study the node voltage sweeps.

Select Remove_trace, Enter, All, and Enter.
Exit the Remove_trace screen.

Current Measurements

Select Add_trace, F4, arrow to I(R1), and Enter twice.

The screen displays the current sweep, I(R1), as the input voltage, Vpwr, is swept. Vpwr is displayed on the X_axis and current, 0A - 10mA, is on the Y_axis. Now, edit a title for the Y_axis.

Select Y_axis, and Enter.
Select Change_title, and Enter.

At Change title: type Circuit Current , and Enter. The Y_axis is titled. Study the display.

Exit the Y_axis screen.
Select Exit, and leave Probe's initial menu screen.

Select Exit_program, and Enter, at the Probe Graphical Waveform Analyzer title screen.

The Control Shell should be displayed.

Browse Output

Page through the data tables. Observe the sweep steps of circuit values that were requested in .print statements. Probe was using this information, converted into a data file (.DAT), to present the "O'scope" graphs that we were observing. Press Esc.

Select Quit, and Exit to DOS, and Enter.
Accept (S) for save changes to the SERIES1.CIR file, and Enter.

Conclusion of the SERIES1 section.

We will use PSpice to analyze branch currents in the next section of the chapter.

II. Branch Currents with Two Voltage Sources

PSpice's adeptness at simulating circuits with more than one voltage source is considered in the following exercise. All circuit voltages and currents are first calculated for expected values.

The Test Circuit

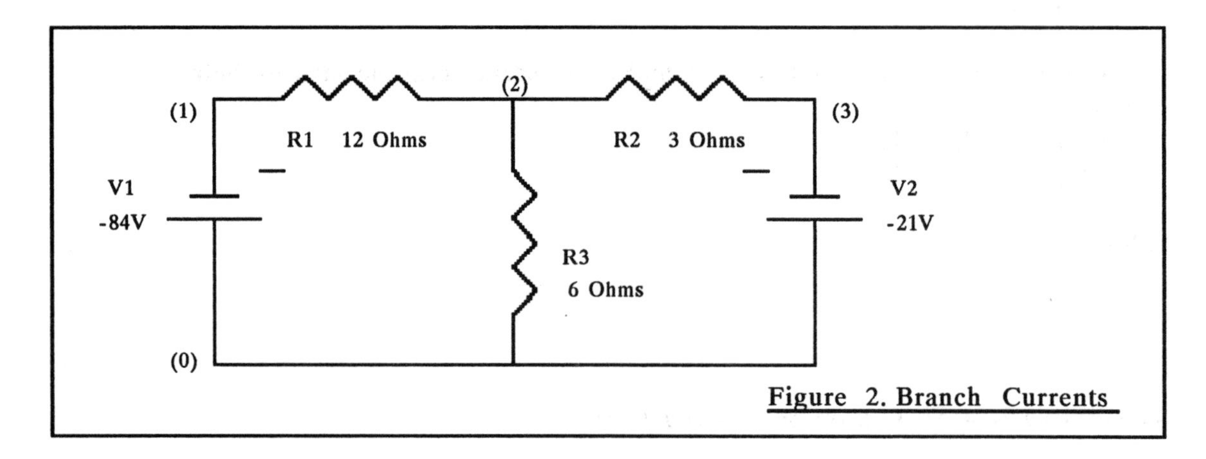

Figure 2. Branch Currents

Circuit Calculations

Branch current problems with more than one source are familiar to beginning electronics students. An appreciation for quick PSpice simulation and analysis will be developed after solving the following lengthy loop simultaneous equations. It is for this reason that all calculations are presented. Without a circuit simulator (or an appropriate network theorem) the solutions can be quite involved. Branch equations are written and solved for loop currents I1 and I2, the currents through V1 and V2. Current through R3 is found by subtracting one source current from the other.

Loop equations are written for each source assuming electron flow clockwise movement of current in Loop 1 and counter-clockwise movement in Loop 2:

84 - VR1 - VR3 = 0V Loop 1
21 - VR2 - VR3 = 0V Loop 2

Using the known resistance values, the equations expand to:

84 - 12(I1) - 6(I1 + I2) = 0V Loop 1
21 - 3(I2) - 6(I1 + I2) = 0V Loop 2

Multiplying the current values in parenthesis by 6, and combining terms, the equations can be written:

-18(I1) - 6(I2) = -84
- 6(I1) - 9(I2) = -21

Dividing the first equation by -6 and the second by -3, we have:

3(I1) + I2 = 14
2(I1) + 3(I2) = 7

Multiply the first equation by 3 to make the second term the same in both equations:

9(I1) + 3(I2) = 42
2(I1) + 3(I2) = 7

Subtract the second equation:

7(I1) = 35
I1 = 5A

Substituting I1 = 5A into the second equation:

2(5) + 3(I2) = 7
I2 = -1A (Minus indicates current opposite the assumed direction, and opposite from I1 through R3)

And finally,

IR3 = I1 + I2
= 5 + (-1)
= 4A

The resistor voltages are calculated using the currents:

VR1 = 5A x 12 Ohms = 60 V
VR2 = 1A x 3 Ohms = 3 V
VR3 = 4A x 6 Ohms = 24 V

All node voltages in PSpice have a Node 0 reference. Both sources are negative in this circuit.

Node 1 = V1 = -84V
Node 2 = VR3 = -24V
Node 3 = V2 = -21V

PSpice Simulation and Analysis

Enter the following netlist with the Circuit Editor.
Filename: BRANCHI1

The Netlist

```
*BRANCH CURRENTS #1
V1 1 0 DC -84V
V2 3 0 DC -21V
R1 1 2 12Ohms
R2 2 3  3Ohms
R3 2 0  6Ohms
.End
```

Press Esc, and Enter, to save the circuit file.

No other control statements are necessary since PSpice automatically prints node voltages and source currents in the Output file (BRANCHI1.OUT).

Run PSpice

Select Analysis, Enter, Run PSpice, and Enter.

Browse Output

Select Files, Enter, Browse Output, and Enter.

Browse the Output file. Examine node voltages and source currents and compare with circuit calculations. They should be the same.

Note: Total Power Dissipation of 399 Watts by PSpice is a result
of adding the -21 Watts in V2. (PV2 = -1A x 21V = -21W)
V1 power dissipation is, PV1 = IV1 x V1 = 5A x 84V = 420W.

Exit Browse Output.

III. Independent Current Sources

The voltage source, Vpwr, used in the series circuit in this chapter is called an independent voltage source. Current sources are also independent sources in PSpice and are written, for examples:

```
I(name)  +Node  -Node  (DC value)  (AC (sweep) value)  (transient values)
I2   6  0  DC  250mA                            ;dc source,  .25A or 250E-3
Iac  4  0  SIN(0  .05A  1K  0  0  0)  ;ac, transient values: sine wave, 50mA, 1KHz
```

In the second statement, the current source, I2, is connected at Node 6, positive node, and Node 0, negative node, and is a DC source of 250 milliamps. But we need to consider the polarity of voltages in the circuit when using current sources. With conventional flow, the current inside the current source flows from the positive node through the source to the negative node. Therefore, current enters the source at the positive node and exits at the negative node. This direction is opposite the voltage source current used earlier where current flowed from positive to negative through the circuit. Since current flow is in the opposite direction, then all voltage polarities in the circuit will be reversed also. To reverse the flow of current when using current sources, simply reverse source node connections or change source polarity, i.e., -10mA in the following example.

Using the series circuit on page 6-2, insert a 10mA current source.
Filename: SERIES1

```
*Series Circuit #1, Circuit description
Is   1 0 DC 10mA
R1  1 2 500
R2  2 3 1000
R3  3 0 1500
.End
```

Run PSpice

Select Analysis, Enter, Run PSpice, and Enter.

Are node voltages V(1), V(2), and V(3) the same as with the voltage source, Vpwr? (p.6-5) The current is the same, 10mA, therefore, the node voltages are the same, but with one exception...the polarities are negative. (Reverse Is node connections later and Run PSpice to check voltage polarities)

Insert the following control statements to sweep the current source, Is, and to measure the voltages and current specified.

```
.DC Is .01 .01 .001                  ;sweep Is___.01A to .01A in .001A steps
.Print DC V(1,2) V(2,3) V(3,0)  ;same as page 6-5
.Print DC I(R1)                       ;same as page 6-5
```

Run PSpice

Are the circuit values the same as when sweeping the voltage source? (p.6-6) The current sweep is the same as when the voltage source was swept in the previous experiment, and all measurements should be the same...except the polarities are reversed. Change the polarity of the source (-10mA) and .DC sweep quantities (-.01 -.01 .001), then Run PSpice to check voltage polarities.

AC Current Sources

Remove the DC current source, .DC sweep statement and .print statements from the series circuit netlist and insert the following AC current input statement and transient analysis control statement. The transient analysis statement is needed to oserve the 1KHz AC waveform and is discussed in other chapters.

```
Iac  1  0  SIN(0  .01A  1K  0  0  0)    ;sine wave input, 10mA, 1KHz
.Tran  .02m  2m  0  .01m               ;final time= 2ms, for analysis  of 2Hz at 1KHz
```

Run PSpice

At Probe, use Add_trace and display I(Iac) and I(R1). Notice the input amplitude has a peak value of 10mA as defined in the input statement. There is a 180 degree phase difference between the two currents. I(Iac) is current through the source. I(R1) is circuit current. Remove both traces.

Select Add_trace and display V(1), V(2) and V(3). Node voltages are in peak values and waveforms are, of course, in-phase with circuit current. Are the node voltages proportional to the resistive divider?

Exit Probe and return to the Control Shell.

Stimulus Editor

Use the Stimulus Editor to observe Iac, the input current waveform. Select Modify_stimulus. Iac is the stimulus to the series circuit and its parameters can be examined and edited. Select Transient_parameters. IOFF, IAMPL, FREQ, TD, DF, and PHASE are available for adjustment. Exit to StmEd initial menu screen.

Another independent current source can be edited into the circuit netlist by selecting New_stimulus. Try this option and set IAMPL= .01A and FREQ= 1500. Return to the Circuit Editor and examine StmEd netlist editing.

Erase the new current source netlist line edited by StmEd and exit PSpice.

Conclusion of Chapter 6

Use "The Scientific Method in PSpice Experimentation" when creating exercises and solving problems. (p.iv)

EOC Problems

(1) Repeat the series circuit exercise with other source and component values. Calculate all circuit expectations before running PSpice.

(2) Repeat the branch currents experiment with other sources and component values. Always calculate, then Run PSpice to verify data.

(3) Using the circuit on page 6-8, remove the voltage sources and edit in current sources to match the currents in the original circuit. Run PSpice to verify voltages are the same. New sources lines are I1 1 0 5A and I2 3 0 -1A, or I2 0 3 1A.

Chapter 7

THEVENIN AND NORTON

Component Models: RES CAP IND

Maximum Power Transfer: Impedance Matching

Objectives: (1) To Thevenize/Nortonize with PSpice
(2) To introduce component models
(3) To illustrate maximum power transfer

I. Thevenin and PSpice

Thevenin's theorem, that an entire resistive network can be represented by a single voltage in series with a single resistance, has challenged beginning electronics students for years. But to PSpice it is one of those circuit challenges that is easily simplified. Textbook and PSpice approaches are the same for finding Thevenin voltage but PSpice uses a variable resistor model to find Thevenin resistance.

Textbook Approach

(1) Disconnect the "load" at terminals A and B.
(Any two points of interest in the circuit)
(2) Calculate or measure the voltage at terminals A and B.
This is Thevenin voltage, Vth.
(3) Disconnect the voltage source and short the source nodes.
(4) Calculate or measure the resistance at terminals A and B.
This is Thevenin resistance, Rth.

PSpice Approach

The Test Circuit

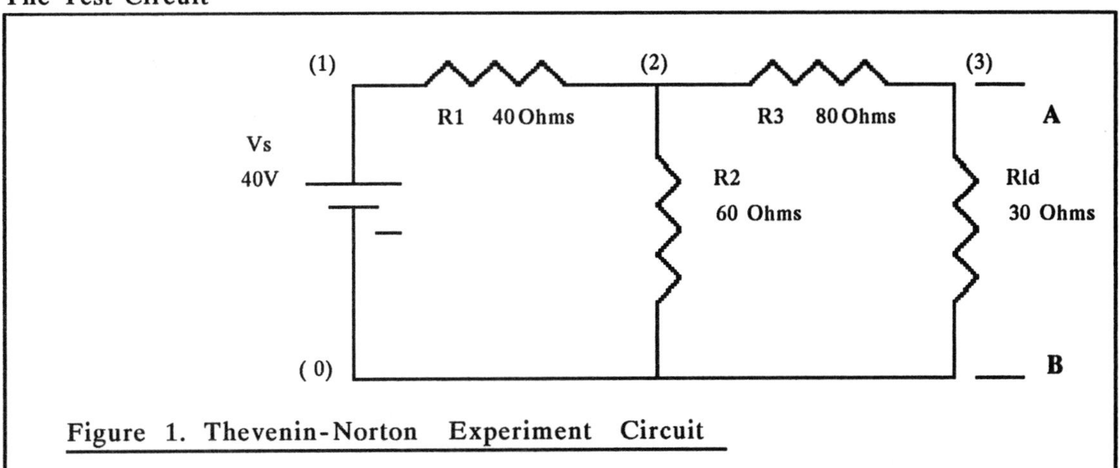

Figure 1. Thevenin-Norton Experiment Circuit

The Netlist, or PSpice Circuit File

Enter the following netlist using the Circuit Editor. Asterisks (*) suspends lines from the file by making them comments. Those lines suspended are not needed in this first PSpice measurement for Thevenin voltage, Vth.

At the PSEVAL52 directory, type PS, and Enter.
Select Files, and Enter.
Select Current File, and Enter

Type filename: THEVENIN, and Enter.
Select Edit, and Enter.

```
*THEVENIN'S THEOREM MEASUREMENTS
Vs  1 0 DC 40V
R1  1 2 40Ohms
R2  2 0 60Ohms
*R3  2 3 80Ohms
*Rld 3 0 30Ohms
*.DC Vs 40 40 1        ;not needed for quick Vth measurement
.Options Nopage
.End
```

Netlist Comments

Ohms can be omitted from the circuit description statements. PSpice uses ohms automatically with R, henrys with L, and farads with C.

.Options Nopage
This control statement groups the Output file into a continues scroll-type presentation instead of separate pages for each section of the report.

Output File Reporting

SMALL SIGNAL BIAS SOLUTION
PSpice will automatically print all node voltages, source current, and power dissipation in the Output File...if a .print statement is not included. .Print will limit outputs to source and requested data.

A QUICK THEVENIN VOLTAGE

A quick approach to finding Thevenin voltage is to examine the circuit for other nodes where this voltage is found after disconnecting Rld. Then, program PSpice to print that node voltage. This is Node 2 in our circuit, therefore, V(2) is the Thevenin voltage.

In order to measure V(2), we have disconnected Rld and R3 from the netlist by placing an "*" at the beginning of the lines. The lines are now comments and are overlooked by PSpice. The .DC sweep line is not needed for this quick measurement and is also made a comment.

Note: The reason for removing R3 is that PSpice requires a node have at least two circuit connections...no loose or dangling nodes. With Rld removed, R3 is no longer in the circuit and Node 3 has only one connection.

Press Esc to leave the Circuit Editor and save the file. Correct any errors, then run PSpice analysis of the circuit in the next steps.

Run PSpice

Select Analysis, and Enter.
Select Run PSpice, and Enter.

Browse the Output File

Select Files, Enter, Browse Output, and Enter.
Use PgDn, PgUp, and Arrow keys to move through the Output file.
Move to the SMALL SIGNAL BIAS SOLUTION page.

V(2)= 24V, and is the Thevenin voltage, Vth.

In some experimentation this method of finding Thevenin voltage may be convenient, but there is a quicker approach.

A QUICKER THEVENIN VOLTAGE

As you recall, the first step in Thevenizing a circuit is "Disconnect the load." Load means the component(s) in the circuit being driven by the Thevenin equivalent circuit. Disconnecting Rld in the test circuit provides an open circuit at terminals A and B. Another way to achieve this open is to change Rld to a very large resistance. Increasing Rld to 100 megohms will provide a quicker Thevenin voltage measurement.

Edit the netlist to include all lines by removing the asterisks from the R3, Rld, and .DC sweep lines.

Edit the Rld line to Rld 3 0 100E6, for 100 megohms.
Insert the following .print statement: .Print DC V(3)

Node 3 is terminal A, therefore, V(3)= Vth.

The analysis could have been run without .DC sweep, of course, and Node 3 voltage would have been the same, but using the sweep statement allows experience in sweep commands and the selection of specific measurements, in this instance....print V(3). We can print the specific data we need when using the .DC sweep statement. This is important in larger circuits.

Run PSpice

Note: The Control Shell program automatically defaults to a Probe run when an analysis is requested. The program will beep at you with what appears to be error messages from Probe. Everything is A-OK. You are just being informed that there is insufficient data to run Probe. The sweep is constant...Probe needs variables.

Browse Output

Press PgDn to "DC TRANSFER CURVES TEMPERATURE= 27 DEG C"
V(3) is the Thevenin voltage = 24V.

Run through these two procedures several times to gain proficiency in finding Thevenin voltage and in operating the program.

The Thevenin Resistance

Now that we have Thevenin voltage, we can use a measurement that has been around for a long time to find Thevenin resistance. Insert a resistance for Rld that reduces the voltage at terminals A-B to one-half the Thevenin voltage of 24Volts. The inserted resistance equals the resistance of the circuit without Rld and is the Thevenin resistance. But how do we know what resistance to insert? The answer...we make an estimate, then use a resistor model to sweep that estimate. First, mentally short the voltage source nodes and estimate the total resistance of the circuit at A-B without Rld. Next, using a resistor model, insert a slightly larger variable resistor at terminals A and B, and let PSpice step through its resistance range. Estimated circuit resistance is 100 Ohms, so step the resistance from 80 to 110 Ohms. When we Run PSpice, the Output file will list all values of step resistances and voltages in a table. The resistance that drops one-half the Thevenin voltage, or 12Volts, is the Thevenin resistance.

Note: How exactly does this provide Thevenin resistance? In previous lab experiments, electronics students have used this method to measure the output impedance of voltage generators. First, the output voltage of an unloaded generator source is measured. Then, a variable resistor is placed across the source terminals and adjusted to reduce the output voltage to one-half the unloaded voltage (maximum power transfer point). The generator output impedance must be equal to the setting of the variable resistor since it is dropping the other half of the unloaded voltage.
(See page 7-14)

This PSpice exercise uses the same impedance matching approach to measure Thevenin equivalent resistance of a circuit.

Edit the Netlist:

```
*THEVENIN'S THEOREM MEASUREMENTS
Vs   1 0 DC 40V
R1   1 2 40Ohms
R2   2 0 60Ohms
R3   2 3 80Ohms
RVAR  3 0 RA-B 1            ;the variable resistor at A - B. Multiplier=1
.Model  RA-B  RES           ;resistor model to sweep
.DC RES RA-B (R) 80 110 1   ;DC sweep resistance; estimated circuit
*                            resistance is 100 Ohm. Range: 80 to 110
*                            Ohms in 1 Ohm steps.
.Print DC V(3)
.Options Nopage
.End
```

Netlist Comments

Construction of a Variable Resistor Model and Sweeping Resistor Values

RVAR 3 0 RA-B 1

RVAR is the variable resistor placed in the circuit at Nodes 3 and 0. It could have any Rname. The model name is RA-B, for resistance at A - B. It also could be any name as long as that name is entered in the .model statement and .DC sweep statement.

The "1" is the multiplier. Any multiplier can be used, but the table values will not be exactly representative. For example, if the multiplier is "10", the values inserted into the circuit will be 800 to 1100 Ohms, but the Output table and Probe graph will read the smaller sweep numbers, 80 to 110. The operator must remember the larger value range. (Enter 10 as the multiplier later and Run PSpice and note results)

.Model RA-B RES

This is the model statement. RA-B is the name of the model. RES is the device type. CAP is the device type for capacitor; IND for inductor.

.DC RES RA-B (R) 80 110 1

The .DC control statement defines the sweep of the resistor model, RES RA-B. Specifically, the model values will be swept from 80 to 110 Ohms in 1 Ohm steps. The (R) indicates the function of the model which is a resistor. If a capacitor model, use (C). If an inductor model, use (L).

Run PSpice

Probe

The program defaults to Probe. The X_axis displays the swept resistor values, 80 - 110 Ohms. Select Add_trace, and display V(3). Reading the graph, V(3)= 12V, one-half the Thevenin voltage, when R= 104 Ohms. This is the Thevenin resistance, Rth.

Select Exits and Enter to return to the Control Shell.

Browse Output File

Select Files, Enter, Browse Output, Enter.
PgDn to "DC TRANSFER CURVES TEMPERATURE= 27 DEG C"
PgDn to the R - V(3) table.
PgDn to the V(3) value of 12Volts, one-half the Thevenin voltage.

The resistance that sets Node 3 voltage to one-half the Thevenin voltage is 104 Ohms in the table. This is the Thevenin resistance.

Thevenin Voltage, Vth, = 24V
Thevenin Resistance, Rth, = 104 Ohms

Estimating resistance to sweep in larger, more complex circuits is difficult. If the circuit is sufficiently large, it may be easier to sweep a larger range of resistance in larger steps. Then, repeat the procedure with a smaller range and smaller steps.

II. Norton and PSpice

Norton's theorem asserts that an entire resistive network can be represented by a single current source, In, in parallel with a single resistance, Rn. Use the Thevenin netlist from the previous section for this first Norton current measurement, but edit as listed below. Notice that Rld has been changed to a very small resistance. The comment lines are not needed in this test.

Norton Netlist

```
*NORTON'S THEOREM MEASUREMENTS
Vs  1 0 DC 40V
R1  1 2 40Ohms
R2  2 0 60Ohms
R3  2 3 80Ohms
Rld 3 0 30E-9
*RVAR 3 0 RA-B 1              ;the variable resistor at A - B.
*.Model RA-B RES              ;resistor model to sweep thru
*.DC RES RA-B (R) 80 110 1    ;est. ckt resistance of 100 Ohms.
.DC Vs 40 40 1
.Print DC I(Rld)
.Options Nopage
.End
```

Textbook Approach

(1) Short terminals A and B. (The points of interest in the circuit)
(2) Calculate or measure the current through the short. This is the Norton current, In.

PSpice Approach

Rld, the load resistor at terminals A and B, is changed to a short circuit with the 30E-9 entry. Therefore, the current through the load, I(Rld), is the Norton current.

Run PSpice

Browse the Output File

PgDn to "DC TRANSFER CURVES TEMPERATURE= 27 DEG C"

I(Rld) = .2308 A, the Norton current, In.

Norton Resistance

Place asterisks before the .DC Vs 40 40 1 and Rld 3 0 30E-9 statements. Remove asterisks from the other circuit statements. Edit the print statement to .Print DC I(RVAR). The resistor model RA-B has been inserted into the circuit to replace Rld. PSpice will sweep the resistance as listed in the .DC sweep statement, 80 to 110 Ohms, and a table of R vs I(RVAR) values will be placed in the Output file. The resistance value, R, when I(RVAR) is one-half the Norton current of .231A, is the Norton resistance.

The edited netlist:

```
*NORTON'S THEOREM MEASUREMENTS
Vs 1 0 40V
R1 1 2 40Ohms
R2 2 0 60Ohms
R3 2 3 80Ohms
*Rld 3 0 30E-9
RVAR 3 0 RA-B 1
.Model RA-B Res
.DC Res RA-B (R) 80 110 1
*.DC Vs 40 40 1
.Print DC I(RVAR)
.Options Nopage
.End
```

Run PSpice

Probe

Use Add_trace to display I(RVAR).

Reading the graph, I(RVAR) is one-half the Norton current (115.5mA) at approximately 104 Ohms. Rn= 104 Ohms.

Exit Probe

Browse the Output File:

Select Files, Enter, Browse Output, and Enter.
PgDn to "DC TRANSFER CURVES TEMPERATURE= 27 DEG C"
PgDn to the R --- I(RVAR) table.

One-half the Norton current is 115.5mA.
The resistance that reduces I(RVAR) to this value is 104 Ohms.

Norton current, In = .231A
Norton resistance, Rn = 104 Ohms

Exit Browse Output
Exit PSpice

Conclusion of Norton and PSpice

We will take a look at component modeling in the following section.

III. COMPONENT MODELS: RES, CAP, and IND

A resistor model was used in this chapter to sweep a range of values. In the experiment, the model simulated a variable resistor in a circuit. A .DC sweep was used to increment model size while measuring circuit response. We will review resistor modeling in this section and take a look at capacitor and inductor models.

Resistor, capacitor and inductor notations have been used in other chapters to insert fixed components into circuits using the same general format. Examples are:

```
R22  6  0  1.4K     ;1.4 KOhms
C5  12  6  10u      ;10 microfarads
L6  14  21  220m    ;220 millihenries
```

Resistor Models

The Test Circuit

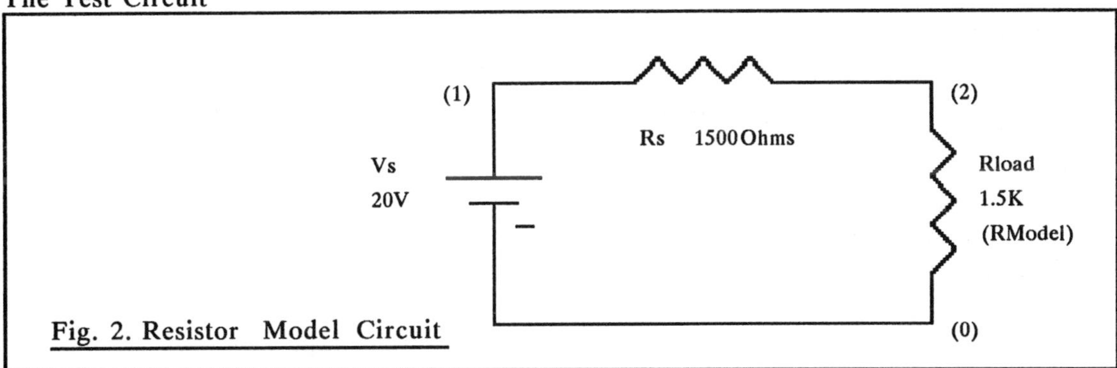

Fig. 2. Resistor Model Circuit

Resistor description statements include the following notations:

Rname +Node -Node {model name} value {temperature coefficient}

Temperature coefficients can be listed here or in .model statements.
Model statements include:

.Model {model name} RES {parameter=value}

Parameters include the multiplier, i.e., R=10, linear temperature coefficient, TC1, quadratic temperature coefficient, TC2, and exponential temperature coefficient, TCE. Default value for the multiplier is 1, and 0 for all temperature coefficients.

Referencing Fixed Resistor Models

Resistor models can be written to be a reference for any resistor description line in the netlist. For example,

```
Rload 2 0 RModel 1.5K
.Model Rmodel RES (R=10)  ;multiplier=X10
```

Parameters in the .model statement are applied to all resistors referencing to that model. For example, a resistor description line that references to .Model RModel RES (R=10) will have its value multiplied by 10. In this example, Rload= 15 KOhms.

Resistor Models and .DC Sweep

Again, to use a resistor model, the component description line must reference to the model name as in the previous example. (The same applies to capacitor and inductor modeling) But if .DC sweep is to be used to test the circuit over a range of resistance values, the description and model lines can take a slightly different form. For example, the listing for resistor and model in the test circuit will include the following lines:

```
Rload  2  0  RModel  1        ;multiplier=X1
.Model  RModel  RES
```

No parameters are specified, therefore, PSpice default values will be used.

Next, we will need a .DC sweep statement to step the resistance through the range of values needed for testing. Probe and the Output file will be used to examine data resulting from the sweep.

```
.DC  RES  RModel  (R)  100  1500  100
```

The .DC sweep must state the type of sweep, RES, and the model to use, RModel in the example. (R) is for resistor device, and the values to sweep are listed, 100 to 1500 Ohms, in 100 Ohms steps. A .Print and .Plot statement will be used to record voltage and current during the sweep.

Filename: RModel

Netlist for resistor model experiment:

```
*RModel Test
Vs  1  0  DC  20V
Rs  1  2  1500
Rload  2  0  RModel  1      ;multiplier=X1
.Model  RModel  RES
.DC  RES  RModel  (R)  100  1500  100
.Print  DC  I(Rload)  V(2)
.Plot  DC  I(Rload)
.End
```

Circuit Expectations

Complete calculations for I(Rload) and V(2) minimum/maximum values.

Sweep	I(Rload)	V(2)
Beginning	____mA	____V
End	____mA	____V

Run PSpice

Select Analysis, Enter, Run PSpice, and Enter.
Observe I(Rload) and V(2) with Probe and in the Output file table and plot.
Compare expected values with PSpice data.

Set Multiplier=10

Change the multiplier to 10, and measure circuit response to the higher range of resistance; 1K - 15K. Note that R values in Probe and in the Output file will be listed as 100 - 1500. The actual resistance values are 1K - 15K, using the X10 multiplier. Step value is 1000.

Complete calculations for I(Rload) and V(2) minimum and maximum values, then Run PSpice.

Sweep	I(Rload)	V(2)
Beginning	____mA	____V
End	____mA	____V

Select Add_trace, and display I(Rload). Estimate I(Rload) at the points where .model resistance is 1KOhm and 15KOhms. Compare calculated values.

Select Remove_trace, and clear the screen.

Select Add_trace, and display V(2). Estimate V(2) where .model resistance is 1K and 15KOhms. Compare these values to calculations.

Exit Probe.

Page through the Output file table and record I(Rload) and V(2) at minimum and maximum values.

I(Rload) = ______Min ______Max

V(2) = ______Min ______Max

Examine the R - I(Rload) plot for I(Rload) =______Min, ______Max.
Compare PSpice data with calculations.
Exit Browse Output.

Capacitor Models: CAP

Capacitor model notations use the same format as resistor modeling with some added features. The circuit description line must, as with the resistor model, call for the model. Initial voltage is an added value that can be entered with capacitors, compared to initial current in inductors. Both provide estimated values to PSpice to use during bias point calculations.

Description line:

```
Cname +Node -Node {model name} value {IC=initial voltage}
```

Initial voltage can be set here or as noted below.

Model statements include:

```
.Model  {model name}  CAP  {parameter=value}
```

Capacitor model parameters include the multiplier, C= , linear voltage coefficient, VC1, quadratic voltage coefficient, VC2, linear temperature coefficient, TC1, and quadratic temperature coefficient, TC2. All default values are 0, except for the multiplier which is 1.

Again, component description lines must reference to the model name as with other modeling. An example is:

```
Cin   4  5  CModel   1u
.Model   CModel   CAP   (C=10)  IC=5V        ;multiplier=10,    IC= 5Volts
```

PSpice has two other methods of setting up the "initial condition" for capacitor models, as well as other initial circuit voltage requirements. One is the .IC statement (Initial bias point condition) for setting initial node voltage. The other is the .Nodeset statement which provides PSpice with an initial node voltage estimate. These settings, like initial voltage parameter values, are held only during the first part of PSpice simulation. .IC settings hold for bias point calculations only. .Nodeset values are abandoned after bias point calculations and the initial step in .DC sweep. In case of duplication in initial settings, .IC overides .Nodeset.

Inductor Models: IND

Inductor model notations use the same format as resistor and capacitor modeling. The circuit description line must, as with the other models, call for the inductor model. Initial current is an added value that is used with inductors, like initial voltage for capacitors. Both provide estimated values to PSpice to use during bias point calculations.

Description line:

Lname +Node -Node {model name} value {IC=initial current}

Model statements include:

.Model {model name} IND {parameter=value}

Inductor model parameters include the multiplier, L= , linear current coefficient, IL1, quadratic current coefficient, IL2, linear temperature coefficient, TC1, and quadratic temperature coefficient, TC2. All default values are 0, except for the multiplier which is 1.

An example of a component description lines that references to the model name is:

```
L14  5  6  LModel  1m
.Model  LModel  IND  (L=1)  IC=.04        ;multiplier=1,  IC=40mA
```

PSpice also uses the .IC statement for setting up the "initial current" for inductors.

Conclusion of Chapter 7

Note: Use "The Scientific Method in PSpice Experimentation" when creating exercises and solving problems. (p.iv) This approach reinforces learning by requiring detailed thought and involvement.

EOC Problems

(1) Write and Run PSpice analysis on the Thevenin equivalent circuit in this chapter; Vth= 24V, Rth= 104 Ohms. Rld= 30 Ohms. Compare original circuit calculations with Thevenin equivalent circuit data.
(2) Write and Run PSpice analysis on the Norton equivalent circuit in this chapter, In= .231A, Rn= 104 Ohms. Rld= 30 Ohms. Compare original circuit calculations with Norton equivalent circuit data.
(3) In the following circuit, find Thevenin voltage and resistance using PSpice methods described in this chapter. The circuit has four series 300 Ohm resistors in parallel with three series 250 Ohm resistors. You provide the voltage source. Draw the circuit, selecting and editing points A and B, and write the netlist.
(4) Using the circuit in Problem 3, find Norton current and resistance using PSpice methods described in this chapter.
(5) Construct several circuits referencing fixed resistor models as defined in this chapter. Run PSpice to verify your calculations.

IV. MAXIMUM POWER TRANSFER MEASUREMENT

In this exercise, we will use PSpice to illustrate impedance matching in maximum power transfer measurements as defined in the note on page 7-4. The voltage generator is connected to a resistor model that will be varied in size with a .DC sweep statement.

Source voltage, Vs= 10V.
Source resistance, Rs= 1.8KOhms.

We will use resistor modeling and .DC sweep information from this chapter to experiment with a variable resistor at Node 2.

Sketch the circuit and complete all calculations before running PSpice analysis. Use several estimates for Rld to prove maximum power is transfered when the impedance of Rld and the source, Rs, are the same.

Enter the following netlist with the Circuit Editor.
Netlist filename: MAXPOWER

```
*MAXIMUM POWER TRANSFER EXPERIMENT
Vs 1 0 DC 10V
Rs 1 2 1.8K
Rld 2 0 RLoad 10        ;multiplier=10
.Model RLoad RES
.DC RES RLoad (R) 100 400 10      ;sweep: 1K - 4K
.End
```

Press Esc to save the circuit file.

Run PSpice

Probe

Select Add_trace, and Enter.
Type V(2)*I(Rld) , and Enter.
Note the "*" in the above multiplication.

PSpice will plot the power equation, P = IV, for power delivered to the load, Rld, during the .DC resistance sweep. Maximum power transfer is indicated at the point where Rld equals the source resistance, 1.8KOhms.

Maximum power = ______mW.
Compare your calculations with PSpice data.

Exercise:

(1) Design several experiments measuring maximum power transfer using this impedance matching example.

Chapter 8

FULL-WAVE RECTIFIER BY PSPICE

BRIDGE RECTIFIER BY PSPICE

Using Transformers in PSpice

Objectives: (1) To PSpice a Full-Wave Rectifier
(2) To PSpice a Bridge Rectifier
(3) To use Probe to analyze rectifier waveforms
(4) To introduce transformer Spiceing

I. Full-Wave Rectifiers by PSpice

Converting power company alternating current to direct current for use in electronic equipment is achieved with rectifiers. Diodes placed in circuits switch on and off in response to incoming bias voltages. Usually, transformers are used to set the output voltage to desired levels. In this exercise, the design of full-wave rectifiers will be evaluated. PSpice will simulate and analyze the design and Probe will display circuit outputs. But first, we need to know what to expect from the rectifier circuit. (Ref:Electronics I theory text)

The Test Circuit

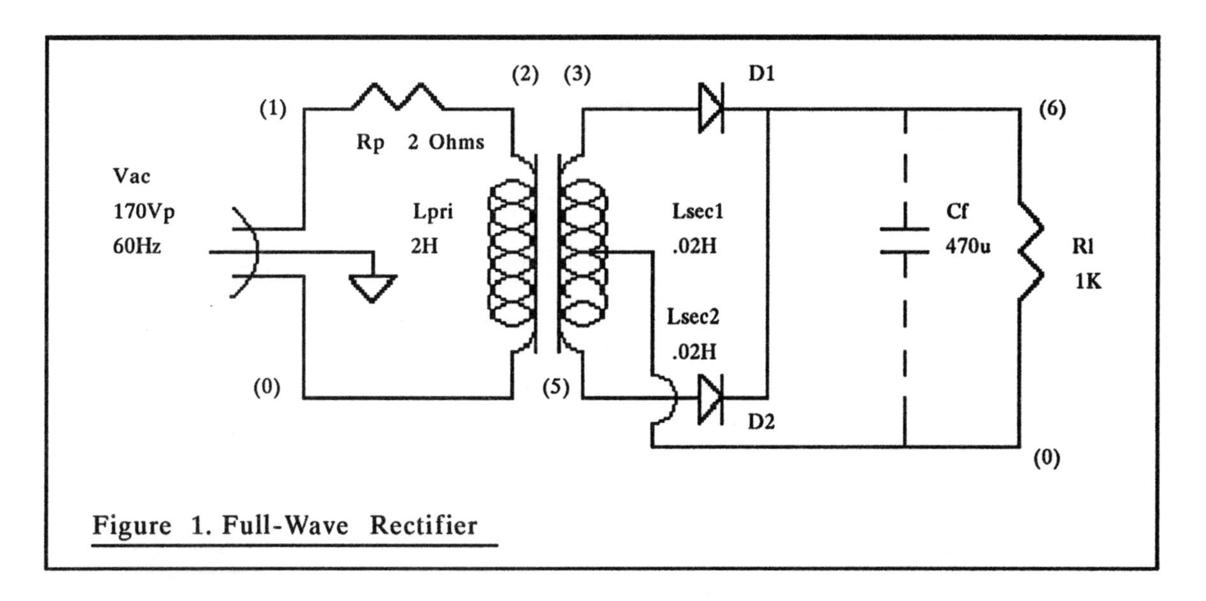

Figure 1. Full-Wave Rectifier

The Circuit File: NETLIST

Enter the following netlist with the Circuit Editor.

Filename: FWRECT, for Full-Wave Rectifier.

```
*FULL-WAVE RECTIFIER
Vac 1 0 SIN(0 170V 60Hz)
Rp  1 2 2Ohm
Lpri 2 0 2H
Lsec1 3 0 .02H
Lsec2 0 5 .02H
KTRANFRM Lpri Lsec1 Lsec2 .999
D1 3 6 D1N4001
D2 5 6 D1N4001
*Cf 6 0 470u
R1 6 0 1K
.TRAN .33m 33.3m 0 .1m
.Model D1N4001 D(Is=10.0E-15 Rs=.1 Ikf=0 Cjo=1p N=1
+              Eg=1.11 Xti=3 Vj=.75 Fc=.5 Nr=2 Bv=100
+              Ibv=100.0E-6 Tt=5n Isr=100.0E-12)
.Options Nopage
.Probe
.End
```

Press Esc to exit the Circuit Editor and save the circuit file.

Netlist Notes

Vac 1 0 SIN(0 170V 60hz) describes the power company line voltage, 170Vpeak. (Vp=Vrms/.707) The input is a 60Hz sine wave.

Rp 1 2 2Ohms. PSpice will not allow inductors to be connected in parallel with voltage sources without some resistance in series. Inductors are treated as AC voltage sources and two voltage sources cannot be connected in parallel. Lpri is connected in parallel with the power company source. Rp is breaking the voltage loop.

Lpri 2 0 2H
Lsec1 3 0 .02H
Lsec2 0 5 .02H These lines define the input transformer with the help of the coupling coefficient statement. In this exercise, the output voltage objective is 17Volts. Each section of the secondary is a 10:1 step-down turns ratio. With an assumed primary inductance of 2H, dividing Lpri by the square of the turns ratio (100) gives the inductance of each section of the secondary... .02H. If secondary inductance is assumed, multiply by the square of the turns ratio to solve for primary inductance.

KTRANFRM Lpri Lsec1 Lsec2 .999. Inductor coupling statement. Coupling coefficient is set at .999. See Using Transformers in PSpice on page 8-11.

Cf 6 0 1800u Capacitor-input Filter. The line is made a comment by inserting the "". The "*" will be removed later in the exercise and the capacitor will be part of the circuit.

D2 5 6 D1N4001 - rectifier power diodes.
D1 3 6 D1N4001

```
.Model  D1N4001  D(Is=10.0E-15 Rs=.1 Ikf=0 Cjo=1p N=1
+                  Eg=1.11 Xti=3 Vj=.75 Fc=.5 Nr=2 Bv=100
+                  Ibv=100.0E-6  Tt=5n Isr=100.0E-12)
```

The 1N4001 model is not a part of the EVAL.LIB file. A model must be written. Be sure to include "+" to begin additional lines.

.TRAN .33m 33.3m 0 .1m. This line programs a transient analysis to be performed on the circuit. The time period of 1 cycle of 60Hz is 16.6ms. Final_time is set to 33.3ms for an analysis period of 2 cycles...the analysis run time. Print_step time interval is .33ms for .print and .plot results from transient analysis, and stored for Probe.

The last entry is step_ceiling and is entered at .1ms. It sets the time of the transient analysis internal time step. In PSpice, step_ ceiling default is final_time/50. This would be .66ms for this analysis but the .1ms step_ceiling will provide more evenly drawn Probe graphics.

.Probe tells PSpice to save all data from the analysis for use by the Probe Graphical Waveform Analyzer. If no values are specified after .Probe, all node voltages and device currents will be save in a data file. (An entry like .Probe V(6) I(D1) will limit the data file to these two measurements) Data files have the original circuit filename with a .DAT extension. Extensive data is made available in this file as will be demonstrated with the Probe program.

.Options Nopage will suppress paging and repeating banners in the Output file.

.End indicates the end of the circuit. PSpice simulation and analysis begins only if the circuit file has the .End statement. Another circuit netlist may be entered after .End, but it must also conclude with the .End statement.

Circuit Expectations

In this first exercise, ideal conditions, including diode first approximations, are assumed in circuit calculations. Probe analysis will display the margin of error introduced in assuming ideal estimates. Rectifier designs normally require more detailed approximations.

A 10:1 turns ratio to each winding of the secondary couples a secondary peak voltage,

Vpsec = Vpri / 10
 = 17Vpeak

Average or DC Value is calculated;

Vdc = 17Vpeak x .636
 = 10.8V

Note: Later, when Cf is inserted into the netlist, the estimated DC output voltage is approximately 17V (ideal).

I(Rl) = V(Rl) / Rl = 17V / 1K
 = 17mA

Next, we will run PSpice and take a look at circuit operation using computer analysis. Probe will be used to display voltages and currents throughout the circuit.

Run PSpice

Select Analysis, and Enter.
Select Run PSpice, and Enter.

Probe's initial menu screen will be displayed when the analysis is completed.

Circuit Voltage

Select Add_trace, and Enter.
Press F4 for a selection of circuit voltages and currents to view.
Arrow to V(6), and Enter twice.

The waveform on the screen is the output of the full-wave rectifier. The X_axis time base (35ms) allows the analysis of two cycles of the 60Hz input...four alternations. V(6)= approximately 16.3V. Our ideal calculations are off by approximately .7V, the diode forward bias voltage.

Select Add_trace, and plot V(3). (Secondary winding voltage)
Select Add_trace, and plot V(5). (Secondary winding voltage)
Note the ~.7V difference comparing V(6) to V(3) and V(5) waveforms.

The Probe Graphical Waveform Analyzer presents an excellent view of waveforms within the full-wave rectifier. Study the traces.

Select Add_trace, and plot V(2), the primary voltage.

The importance of computer aided design is dramatically demonstrated in the measurements available on Probe screens. Circuit theory is reinforced by these practical "O'scope" readings. And voltage is just part of the analysis.

Circuit Current

Select Remove_trace, Enter, All, and Enter twice.
Select Add_trace, and plot I(D1), the current through D1.
Repeat for Diode, D2.

Diode current is measures above 16mA peak.

Select Remove_trace, Enter, All, and Enter twice.
Select Add_trace, and plot I(Lsec1) and I(Lsec2).

A slight delay in diode turn-on is indicated by the alternations not joining at the center of the screen.

Insert the Capacitor-Input Filter, Cf

Follow the next steps and return to the Circuit Editor to place the filter capacitor, Cf, into the circuit netlist.

Select Exit, and Enter, to leave the initial Probe screen. The Probe title screen will be displayed.
Select Exit_program, and Enter, to leave Probe and return to the Control Shell.
Select Files, Enter, Edit, and Enter.
Remove the "*" comment insertion and make the Cf line part of the netlist. If properly designed, the capacitor-input filter should reduce ripple voltage to less than 10% of the DC level.
Press Esc to exit and save the circuit file.

Run PSpice

DC Output

Select Add_trace, and plot V(6), the output voltage.

V(6) is approximately 16.3Volts with less than 4% ripple.

Select Remove_trace, and clear the screen.
View I(Rl), the load current.
Load current is approximately 16.3mA.

Surge Current

Select Remove_trace, and clear the screen.

Next, examine the initial surge current through D1.

View I(D1). Initial surge is less than 5A for <4ms.

(In the 1N4001, IFSM= 30A for one cycle, 24A for two cycles)

View I(D2) and I(Cf). Study, then clear the screen.

Examine I(Lpri) for a maximum surge of less than .5A at the beginning of the analysis.

Exit Probe

This concludes the section on full-wave rectifiers using PSpice. Bridge rectifiers will be simulated and analyzed in the next section.

II. Bridge Rectifiers by PSpice

The output of a bridge rectifier is very much the same as that of a full-wave rectifier with one exception. With the same peak secondary voltage the DC output is approximately doubled because of the absence of the center tap in the secondary. With that in mind, and assuming the previous section examining the full-wave rectifier has been completed, let's move directly into the operation of the bridge rectifier.

The Test Circuit

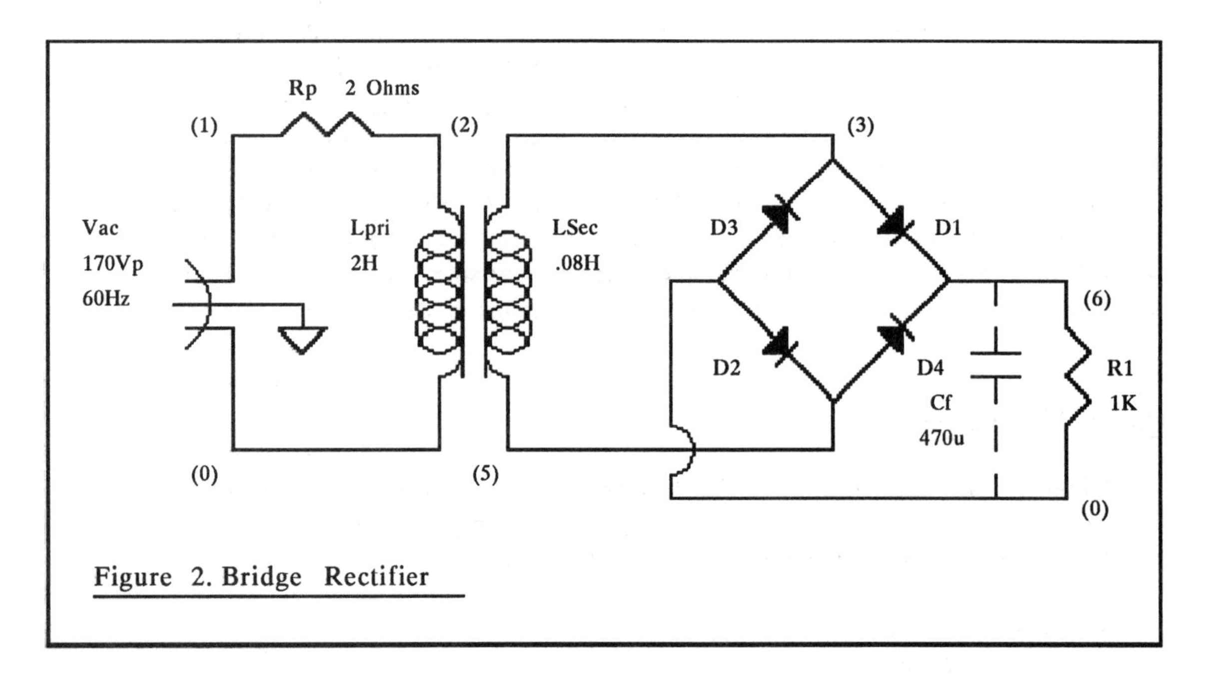

Figure 2. Bridge Rectifier

The Circuit File: NETLIST

Enter the netlist with the Circuit Editor.

Filename: BRECTIFIER, for Bridge Rectifier.

```
*BRIDGE RECTIFIER
Vac  1  0 SIN(0  170V  60Hz)
Rp   1  2  2Ohm
Lpri 2  0  2H
Lsec  3  5  .08H     ;5:1 turns ratio
KTRANFRM  Lpri  Lsec  .999
D1  3  6  D1N4001
D2  5  6  D1N4001
D3  0  3  D1N4001
D4  0  5  D1N4001
*Cf  6  0  470u
R1  6  0  1K
.TRAN  .33m  33.3m  0  .1m
.Model D1N4001 D(Is=10.0E-15 Rs=.1 Ikf=0 Cjo=1p N=1
+                  Eg=1.11 Xti=3 Vj=.75 Fc=.5 Nr=2 Bv=100
+                  Ibv=100.0E-6  Tt=5n  Isr=100.0E-12)
.Options Nopage
.Probe
.End
```

Press Esc to exit the Circuit Editor and save the circuit file.

Netlist Notes

Transformer turns ratio is one of the first considerations when designing rectifiers. Secondary peak voltage, a turns ratio step-up or step-down from input peak voltage, is very near the estimated DC output voltage. Winding inductance values are calculated by the square of turns ratio, 5:1 in this circuit. Assuming the secondary winding value is .08H, multiply by the square of the ratio to obtain the primary value (.08H X 25 = 2H). If the primary winding value is known, divide by the square of the turns ratio (2H / 25 = .08H). Other parameters and limits are, of course, included in transformer design. Examples are current and power requirements, plus core, leakage, losses, and physical specs.

Circuit Expectations

Diode second approximations should improve estimated values. Each diode is forward biased with .7V, for a total of 1.4V drop in output voltage.

A 5:1 turns ratio couples a secondary peak voltage,

Vpsec = Vpri / 5
= 34Vpeak

Output voltage, V(6) = 34Vp - 1.4V
= 32.6Vpeak

Average or DC value of this bridge rectifier output waveform is,

Vdc = (34Vpeak - 1.4V) x .636

= 20.7V

Note: Later, when Cf is inserted into the netlist, the estimated DC output voltage is approximately 32.6V.

I(Rl) = Vload / Rl = 32.6V / 1K

= 32.6mA

Next, we will run PSpice and take a look at circuit operation using computer analysis. Probe will display voltages and currents throughout the circuit.

Run PSpice

Select Analysis, and Enter.
Select Run PSpice, and Enter.

Probe's initial screen will be displayed when PSpice analysis is completed.

Circuit Voltage

Select Add_trace, and Enter.
Press F4 for a selection of circuit voltages and currents to view.
Arrow to V(6), and Enter twice.

The waveform on the screen is the output of the bridge bectifier. The X_axis time base (35ms) allows the analysis of two cycles of the 60Hz input...four alternations. Vpeak is near the estimated 32.6Volts.

Select Add_trace, and plot V(3). (Secondary winding voltage)
Select Add_trace, and plot V(5). (Secondary winding voltage)

The Probe Waveform Analyzer presents an excellent view of waveforms on the secondary windings of the bridge rectifier. The amplitude differences comparing V(6) to V(3) and V(5) include diode forward bias voltage.

Select Add_trace, and plot V(2), the primary voltage.

Again, the importance of computer aided design is dramatically demonstrated. Circuit theory is supported by practical "O'scope" presentations. Next, current waveforms in the rectifier will be displayed.

Circuit Current

Select Remove_trace, Enter, All, and Enter twice.
Select Add_trace, and plot I(D1), the current through D1.
Repeat for Diode, D4.
Repeat for Diodes D2 and D3.
The pulsating DC current level is above 32mA peak. Diodes D1, D4 and D2, D3 conduct simultaneously on opposite sides of the bridge.
Select Remove_trace, Enter, All, and Enter twice.
Select Add_trace, and plot I(Lsec).

A slight delay caused by the required diode forward bias voltage of 1.4V is indicated by the alternations not joining at the center of the display.

Insert the Capacitor-Input Filter, Cf

Follow the next steps and return to the Circuit Editor to place the filter capacitor, Cf, into the circuit netlist.

Select Exit, and Enter, to leave the initial Probe screen. The Probe title screen will be displayed.
Select Exit_program, and Enter, to leave the Probe Graphical Waveform Analyzer.
Select Files, Enter, Edit, and Enter.
Remove the "*" comment insertion and make the Cf line part of the program. If properly designed, the capacitor-input filter will reduce ripple voltage to less than 10% of the DC output voltage.

Run PSpice

DC Output

Select Add_trace, and plot V(6), the DC output voltage.

The rectifier output, V(6), is approximately 32Volts with less than 4% ripple.

Select Remove_trace, and clear the screen.
View I(Rl), the load current.

I(Rl) = 32mA

Next, view the initial surge current through D1.

Select Remove_trace, and clear the screen.
View I(D1). Initial surge= 8.5A maximum for less than 4ms.
(In the 1N4001, I(FSM)= 30A for one cycle)

View I(D3) and I(Cf).

Select Remove_trace, and clear the screen.
View I(Lpri). The primary current is 1.75A maximum for less than 4ms at the beginning of the analysis.

Exit Probe and PSpice

Conclusion of Chapter 8

EOC Problems

(1) Using your theory text as a guide, design a full-wave rectifier with an output of 5Vdc. Write and enter the circuit netlist and run PSpice analysis of your design.

(2) Using your theory text as a guide, design a bridge rectifier with a 5Vdc output. Write and enter the circuit netlist and run PSpice analysis of your design.

(3) Design a half-wave rectifier with a 12Vdc output. Write the netlist and run an analysis of this circuit.

(4) Design a step-up transformer and bridge rectifier to boost 10Vac to 47Vdc. Using your theory text as a guide, design the capacitor input filter for a ripple of exactly 10%.

We will take a look at using transformers in PSpice in the next section.

III. Using Transformers in PSpice

Two types of lines are required when entering a transformer in the netlist. First, the inductors that are transformer windings are listed, and second, the K line, or inductor coupling (transformer core) line, defines the transformer and its coupling coefficient. (Ref: AC Fundamentals theory text; mutual inductance, coupling coefficient, and transformer sense dots)

A transformer winding is entered the same as an inductor, L, for examples;

```
Lpri   4 5 40m                     L3 16 14 2.5H
Lsec1  6 7 1.6m                    L4 18 17 2.5H
Lsec2  7 6 1.6m
```

Next, the inductor coupling statement is listed. This line must include all windings in the transformer and define the coupling coefficient for each.

```
Ktranfmr  Lpri  Lsec1  Lsec2  .999
KTRAN  L3  L4  .5
```

K(name) couples all inductors listed at the same coupling coefficient....999 and .5 for the two transformers listed above.

Winding Voltage Polarity and Current Direction

Voltage "polarity sensitive" circuit design will require proper PSpice node connections. The polarity of voltage in a transformer winding is defined by the node connections in inductor winding description statements.

```
L1  1 0 .4H
L2  2 0 .4H
K12 L1 L2 .999
```

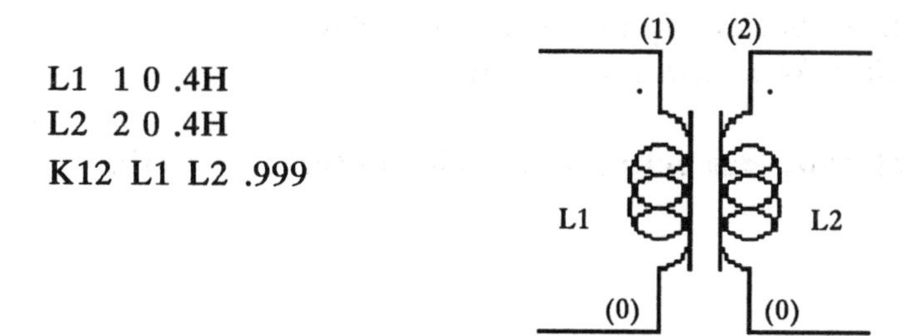

In this diagram, current in the secondary winding, L2, is opposite from current in the primary winding, L1. Voltages at Nodes 1 and 2 are in phase...using the traditional "dot" convention for transformers. In PSpice, a dot is understood to be placed at the first node listed. Reversing node placements reverses voltage polarity.

In the following test circuit, a voltage source will be used to illustrate "dot" in-phase voltages. A current source will be used later for the same purpose and to illustrate current flow in these sources.

Enter the following netlist. Filename: TRANTEST

```
*TRANSFORMER WINDING POLARITY TEST
Vac  4 0 SIN(0  1V  60  0  0  0)
*Iac  4  0 SIN(0  .25A  60  0  0  0)
Rp   4  1  10
L1   1  0  .4H
L2   2  0  .4H
Rld  2  0  10
KTRNX  L1  L2  .999
.Tran  .33m  33.3m  0  .1m
.Probe
.End
```

Sketch a diagram of the circuit and place "dots" at the top of the transformer windings (Nodes 1 and 2). Press Esc and save the file.

Run PSpice

At Probe

Select Add_trace, and display V(1) and V(2). The two voltages are in phase as indicated on the diagram. (Reversing L2 node connections, L2 0 2 .4H, would reverse V(2) polarity and dot notation on the secondary winding would be at Node 0) Remove both traces.

Select Add_trace, and display I(L1) and I(L2). As expected, the primary and secondary currents are flowing in opposite directions. Note polarities and return to the Circuit Editor.

Insert Current Source

Make the voltage source statement a comment, and insert the current source line into the netlist. Sketch the new circuit with the current source. As you recall, current flow in independent current sources is from +Node to -Node, "through the current source." Node 4 will initially have a negative voltage as will be seen in transient analysis. Therefore, initial current in the primary will be opposite from current flow with the independent voltage source. Note this reversing of winding currents using current sources in the next steps.

Run PSpice

At Probe

Select Add_trace, and display V(1) and V(2). The in-phase relationship is the same, but note the negative going waveforms at the beginning of transient analysis. Remove both traces.

Select Add_trace, and display I(L1) and I(L2). Winding currents are out of phase as expected, and opposite from voltage source currents.

Exit Probe and PSpice. Great job!

Chapter 9

BJT COLLECTOR CURVES

JFET DRAIN AND TRANSCONDUCTANCE CURVES

TITLES, LABELS, AND ARROWS IN PROBE

Objectives: (1) To use PSpice to plot BJT collector curves and load lines
(2) To use PSpice to plot JFET drain and transconductance curves
(3) To title and label transistor curve screens using Probe

I. Bipolar Junction Transistor Collector Curves

Transistor DC Beta is the ratio of collector current to base current, IC/IB. Emitter current is the sum of these two currents. Collector curves are plots of base current on a collector current graph. In test setups, collector current (IC) is measured while incrementally changing base current (IB) and sweeping collector-emitter voltage (VCE). A small amount of reverse collector-emitter voltage enables the collector to "collect" all electrons arriving in its depletion area. As VCE increases, the level of collector current is nearly constant for a given base current and rises only slightly until transistor breakdown.

BJT collector curves are separated into four regions that define transistor operation. They are saturation, cutoff, breakdown, and active regions. These regions will be identified and labeled in the following exercise. PSpice will be used to (1) simulate the test setup, (2) to analyze the circuit, (3) to plot base currents on a collector current graph, and (4) to draw typical DC load lines.

BJT Test Circuit

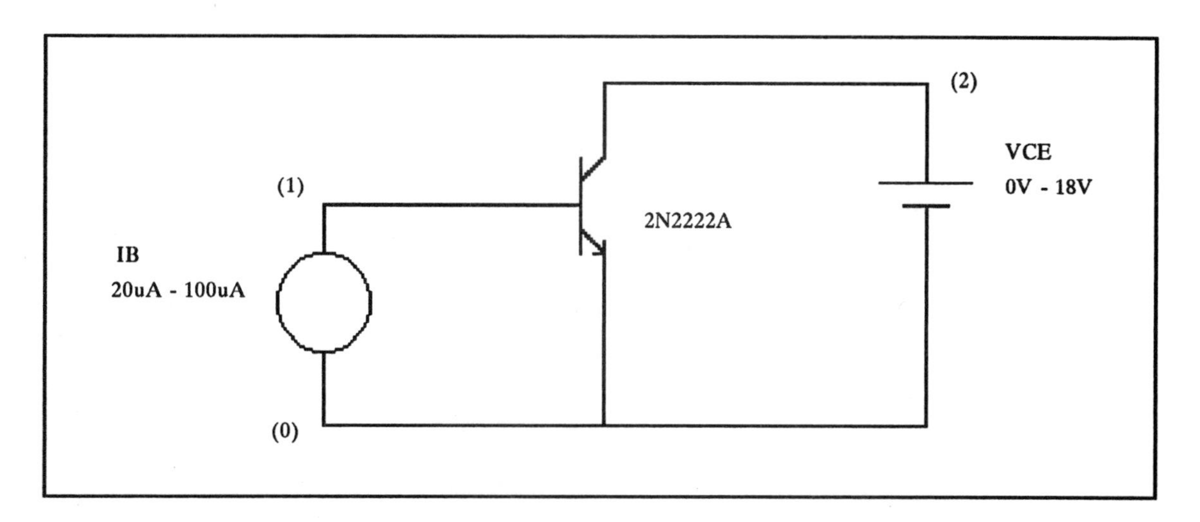

The Circuit File

Enter the following netlist with the Circuit Editor.
Select Files, Enter, Edit, and Enter. Filename: BJTCURVES

```
*BJT COLLECTOR CURVES
VCE  2  0  DC  0
IB     0  1  DC  0
Q1  2  1  0  Q2N2222A
.Model  Q2N2222A   NPN     ;in Evaluation   library
*DC Sweep  Analysis            ;nested  sweep;  VCE, IB
.DC  VCE  0  18  .1  IB  20E-6  100E-6  20E-6 ;begin at 20uA,
*                                                  ;end  at  100uA,  step=20uA.
.LIB  EVAL.LIB
.Probe
.End
```

Press Esc, to exit the Circuit Editor and save the circuit file.

Netlist Notes

Current flow in independent current sources is from +Node to -Node "through the source." Node 0 is the positive node in the IB 0 1 DC 0 statement. Using conventional flow, current flows out of Node 1 and into Node 0. (See Independent Current Sources, Chapter 6)

VCE 2 0 DC 0
IB 0 2 DC 0 These two sweep_variable sources are set to zero since the data of interest will be generated with the values listed in .DC sweep. Collector-emitter voltage and base current are included in the sweep.

.DC VCE 0 18 .1 IB 20E-6 100E-6 20E-6
.DC Sweep statements have been used in this text to sweep DC voltage source values (Chap. 3) and resistor model values (Chap. 7). In this exercise, a nested sweep statement is used to sweep a voltage source while incrementally stepping a current source. VCE will be swept for each value of the second sweep, IB.

PSpice DC analysis begins at a base current of 20uA, and steps 20uA to 100uA. Steps are 20uA, 40uA, 60uA, 80uA, and 100uA. Collector-emitter voltage will sweep from 0V to 18V in .1Volt steps for each value of base current. The progress of the DC analysis is displayed on the screen during the test.

.LIB EVAL.LIB
PSpice will search for the Q2N2222A model in the Evaluation Library. Four BJT models are included in the Evaluation Library; NPNs: Q2N2222A and Q2N3904; PNPs: Q2N3907A and Q2N3906. Examine these models by typing

"Type Eval.Lib" at the PSEVAL52 directory DOS prompt. If you are in Control Shell, you may arrow to Quit, select DOS Command, and press Enter. At "Enter Command", type "Type Eval.Lib" and Enter. Use the Pause key and Spacebar to page through the library. Note the detailed list of BJT model parameter data. The "+" lines are extensions of model definitions and are necessary. The complete model description must be written into the netlist if the model is not available in library files. (Models Q2N2222A BJT and J2N3819 JFET are not exact models in the evaluation version of PSpice)

Follow screen prompts to return to Control Shell.

Ask your instructor for a copy of the Evaluation Library file.

Run PSpice

Select Analysis, and Enter.

Select Run PSpice, and Enter.

PSpice will simulate the test circuit and run the analysis as defined in the netlist. Observe the VCE sweep for each level of IB. The Control Shell program defaults to a Probe run after PSpice analysis. A .Probe statement is included to assure saving a data file for the Probe Graphical Waveform Analyzer program.

At Probe

The following steps will display the collector curves, but first, we need to set Y_axis range. Probe will auto-range the Y_axis to incoming data and we may not get the range needed. Probe will hold a "Set_range."

Select Y_axis, and Enter.

At Y_axis menu, select Set_range, and Enter.

At Enter a range: type 0mA 12mA , and Enter.

The Y_axis scale is set to the values specified.

Select Exit, and Enter.

Collector Curves

Select Add_trace, and Enter.

Press F4 for a selection of variables.

Arrow to IC(Q1), and Enter twice. Note that collector current, IC(Q1), is plotted for the stepped values of base current, IB.

A set of collector curves are calculated by Probe and drawn on the screen. Five collector current lines are drawn equating to the five 20uA base current steps...20uA to 100uA. Collector current is displayed on the Y_axis, and VCE on the X_axis. A DC Beta of 100 is evident on the graph.

TITLES, LABELS, AND ARROWS

Using Arrow Keys

If a Mouse is not available, labeling and drawing arrows is easily accomplished using the Arrow keys. For example, when placing labels, use the Arrow keys to move the label into position, then press Enter. The label is placed. When using Arrow keys to plot an arrow, move the arrow to the start position with the Arrow keys and press Enter. Then, using the Arrow keys, draw the arrow to the desired length and position, and press Enter again. The arrow is placed.

Title Y_axis

Select Y_axis, and Enter.
Select Change_title, and Enter.
At Change title: type COLLECTOR CURRENT, and Enter. The Y_axis is titled "COLLECTOR CURRENT".
Select Exit, and Enter, to leave Y_axis menu.

Labels

Select Label, and Enter.
Select Text, and Enter.
At Enter the Text: type Base Current, and Enter.

Place the label, Base Current, by clicking on the label and dragging to the upper center of the graph, and unclick. If a Mouse is not available, move with the Arrow keys, and press Enter when properly located.

Repeat the last set of steps and label each base current line, beginning at the top: 100uA, 80uA, 60uA, 40uA, and 20uA.

Repeat the steps and label, Saturation Region. Place the label near the upper left corner of the graph. Next, an arrow will be drawn pointing to the satuation region.

Select Arrow, and Enter.

Place the Mouse cursor just beneath the middle of the word Saturation, and click and while holding down the left button, drag the cursor to the saturation region and release. An arrow will be drawn from the label to the saturation region of the collector curves.

Notes: To relocate a label, select "Move" in Label menu and press Enter. Position the mouse cursor on the label and click and drag the label to the new position and unclick.

To remove a label or an arrow, select Delete_one, and Enter. Click on either item to delete it.

Press Esc if menu is not highlighted and restart at Label selection.

Repeat the label steps and place "Breakdown Region" at the upper right corner of the graph. Draw an arrow from the label to the breakdown region.

Repeat the label steps and place "Cutoff Region" at the bottom center of the graph. Draw an arrow from this label to the X_axis base line.

Repeat the label steps and place "Active Region" near the upper center of the curves, just above the 80uA line. Draw arrows to indicate the width of the active region.

Move the Active Region label down to the 60uA line.
Select Move, and Enter.
Select the Active Region label by clicking and holding the left Mouse button. Drag the label down to the 60uA base current line and release.
Move the two arrows that point to the width of the active region.

Place a title label, "2N2222 Transistor Collector Curves", on the 40uA base line, near the middle of the graph.
Select Exit, and Enter, to leave the Label menu screen.

Mark the data points on the graph.
Select Plot_control, and Enter.
Select Mark_data_points, and Enter.
Notice the entire base curve is one data point, as one step in the analysis.
Select Do_not_mark_data_points, and Enter.
Select Exit, and Enter, to leave the Plot_control menu.

Conclusion of using Probe to title and label BJT collector curves.

Drawing Load Lines

The Q-point, or operating point, for a transistor circuit is set on the DC load line. Load lines are drawn in the following steps and are plots of VCE(sweep) minus collector-emitter voltage divided by "load resistance." Load resistance is total DC resistance in saturation calculations.
Select Add trace, and Enter.
Assuming a load resistance of 1500 Ohms.
Type (18-V(2))/1500, and Enter.

Repeat for a load resistance of 2000 Ohms.
Type (18-V(2))/2000, and Enter.

Assuming a fixed VE= 6Volts and a collector resistance of 1500 Ohms, plot (12-V(2))/1500.
Repeat for other values of load resistance.

Sketch Collector Curves

Sketch the collector curves, load lines, and label the graph in Figure 1.

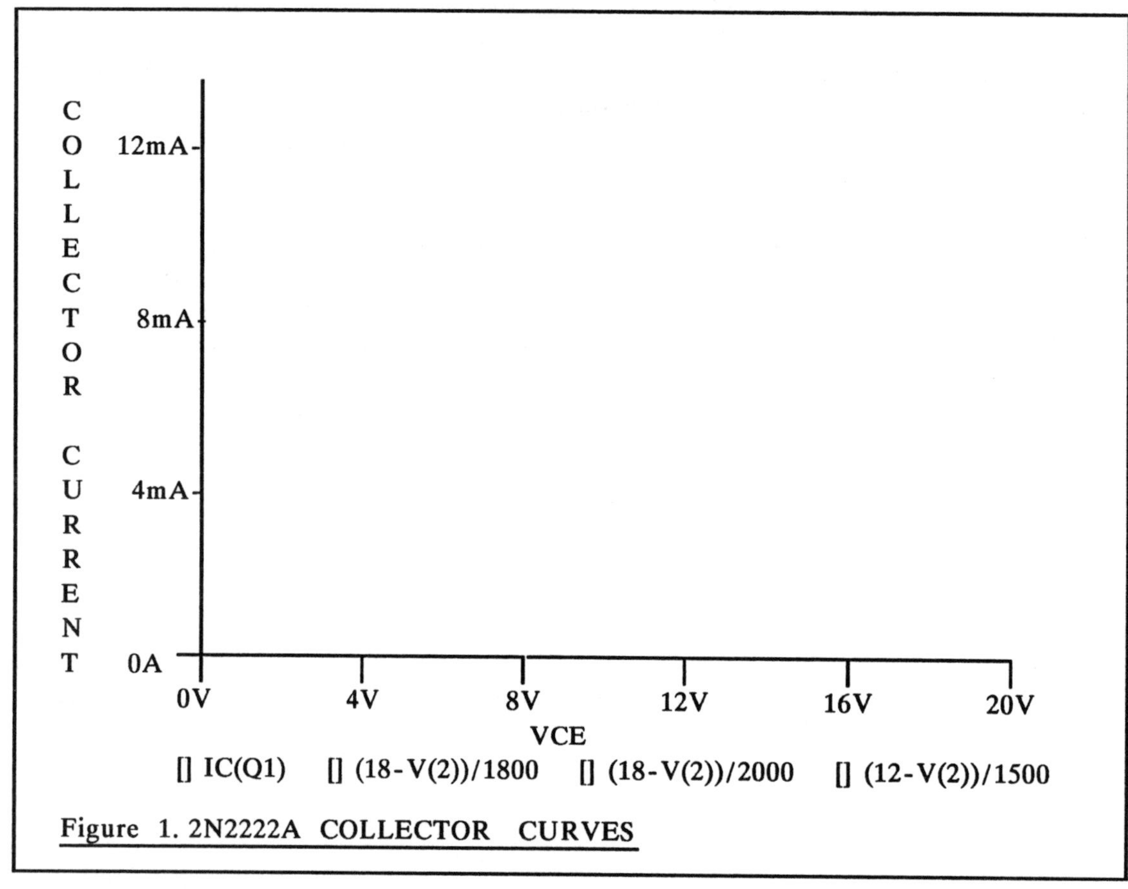

Figure 1. 2N2222A COLLECTOR CURVES

Select Exit, Enter, Exit_program, and Enter to leave Probe.

II. JFET Drain and Transconductance Curves

A set of 2N3819 JFET drain curves, similar to BJT collector curves, will be displayed as a first exercise in this section. A 2N3819 JFET transconductance curve will be analyzed in the last experiment.

The test setup for drain curve measurements is comparable to the one used for collector curves. Gate voltage is incrementally stepped in JFETs instead of base current in BJTs. Drain-source voltage (VDS) is swept and drain current (ID) is measured. To graph these values, drain current is displayed on the Y_axis and VDS is placed along the X_axis.

The operating regions that define JFET operation are ohmic region, active region, gate-source cutoff (or cutoff), and breakdown region. Ohmic region and active (or current-source) region operation requires separate models to define the JFET. Like the BJT, the JFET will operate as a current source along the horizontal active region. The ohmic region equates to the saturation region of the BJT. In this region the JFET operates as a resistor. The pinchoff voltage (defined in earlier studies) separates ohmic and current-source operations. These regions will be observed and labeled in Probe.

It should be noted that these drain curves present ideal operation of the JFET. More refined measurements are required for specific designs. But for general observations and as an excellent learning tool, JFET transistor characteristics are adequately presented.

Two JFET models are included in the evaluation library: J2N3819 and J2N4393. The parameters for each are taken from manufacturer's data sheets. (Model J2N3819 is not an exact model in the evaluation version) Take a look at these models by typing "Type Eval.Lib" at PSEVAL52 directory DOS prompt. If you are at the Control Shell, you may arrow to Quit, select DOS Command, and press Enter. At "Enter Command", type "Type Eval.Lib" and Enter. Use the Pause key and Spacebar to page through the library. Note the detailed list of JFET data. The lines beginning with an asterisk, "*", are comment lines and unnecessary in model definitions. The "+" lines are extensions of definitions and are necessary. Follow the screen prompts to return to Control Shell.

PSpice will be used to simulate the test setup and to analyze the circuit. Test data will be stored in a .DAT file for Probe presentation.

(A) JFET Drain Curves

JFET Test Circuit

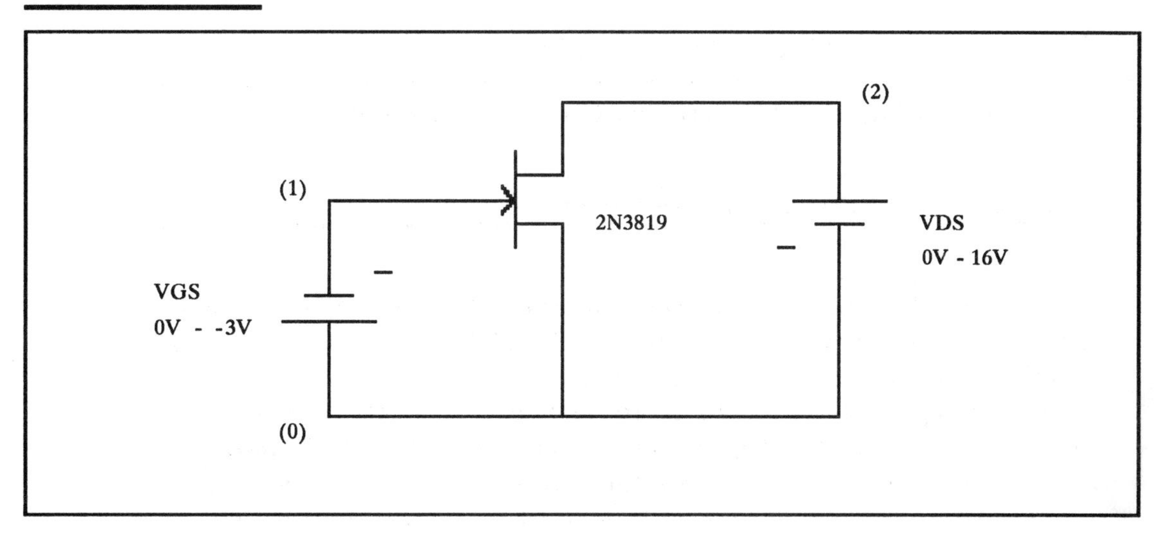

The Netlist

Enter the following netlist at the Circuit Editor.

Filename: JFETDC

```
*JFET DRAIN CURVES, J2N3819
VDS  2 0 DC 0
VGS  1 0 DC 0
J1   2 1 0 J2N3819
.DC  VDS  0  16  .4  VGS  0  -3  .5
.Model J2N3819 NJF(Vto=-3  Is=33.57f  Beta=1.304m)
.LIB  EVAL.LIB
.Probe
.End
```

Press Esc to exit and save the circuit file.

Netlist Notes

VDS 2 0 DC 0
VGS 1 0 DC 0 These two sweep_variable sources are set to zero since the data of interest will be generated with the values listed in .DC sweep. Drain-source voltage and gate-source voltage are part of the sweep statement.

.DC VDS 0 16 .4 VGS 0 -3 .5
In this exercise, a nested_sweep statement is used to sweep a voltage source while incrementally stepping another voltage source. VDS will be swept for each value of the second sweep, VGS.

DC analysis begins at a gate-source voltage of 0V, and steps .5V to -3V. Steps are 0V, -.5V, -1V, -1.5V, -2V, -2.5V, and -3V. Drain-source voltage will sweep from 0V to 16V in .4Volt steps for each value of gate voltage. Drain current is measured during the test and will reach cutoff at the Vto level of -3Volts, the pinchoff voltage.

Observe the monitor during the DC sweep analysis.

.Model J2N3819 NJF(Vto=-3 Is=33.57f Beta=1.304m)
JFET parameters, Vto= -3, Is= 33.57f, and Beta= 1.304m, are copied from the library model for experience only. Adjusting these and all parameters, thereby writing a modified or new model, is accomplished simply by typing in the new data. (See parameter definitions in Appendix B)

Run PSpice

At Probe

Probe's initial menu screen is displayed after PSpice run. Circuit simulation and analysis has been completed and the data from that run is stored in the JFETDC.DAT file for use by Probe.

Select Add_trace, and Enter.

Press F4 for a selection of circuit variables.

Arrow to ID(J1), and press Enter twice.

Drain current curves are displayed beginning at VGS= 0Volts at the top of the graph. Each curve is a -.5Volt step to -3Volts at cutoff.

Reset Y_axis range to allow for a label at the top of the graph.
Select Y_axis, and Enter.
Select Set_range, and Enter.
At Enter a range: type 0mA 14mA , and Enter.

While at the Y_axis menu, enter a title on the Y_axis.
Select Change_title, and Enter.
At Change_title: type Drain Current, and Enter.
Drain Current is the title along the left side of the graph.
At Exit, press Enter.

Labeling the Graph

Select Label, and Enter.
Select Text, and Enter.
At Enter a text: type 2N3819 JFET DRAIN CURVES, and Enter.

The label appears at the center of the screen and must be moved to the top of the graph. With the Mouse, click on the text and drag to the upper center of the graph, and release. If a Mouse is not available, move the label with Arrow keys, and press Enter when properly located.

Next, the gate voltage plots will be labeled.

Select Text, and Enter.
At Enter a text: type VGS= 0V , and Enter.
Drag this label to the upper left corner of the graph.

Repeat for each of the gate voltage plots. The next line down is VGS= -.5V, then VGS= -1V, etc. Place each label just above the line. For the cutoff voltage of VGS= -3V, place the label in the middle right section of the graph and draw an arrow to the cutoff line.

Select Arrow, and Enter. Place the tail of the arrow at VGS= -3V, click and drag the arrow to the gate cutoff voltage line, and release. If Arrow keys are used, move the cursor to VGS= -3V, and press Enter. Move the cursor to the cutoff voltage line and press Enter again.

Draw an arrow from the VGS= 0V label to the top line.

Labeling of the JFET drain curves is completed.

Sketch Drain Curve Plots

Sketch the various gate voltage plots and label the graph in Figure 2 on the next page.

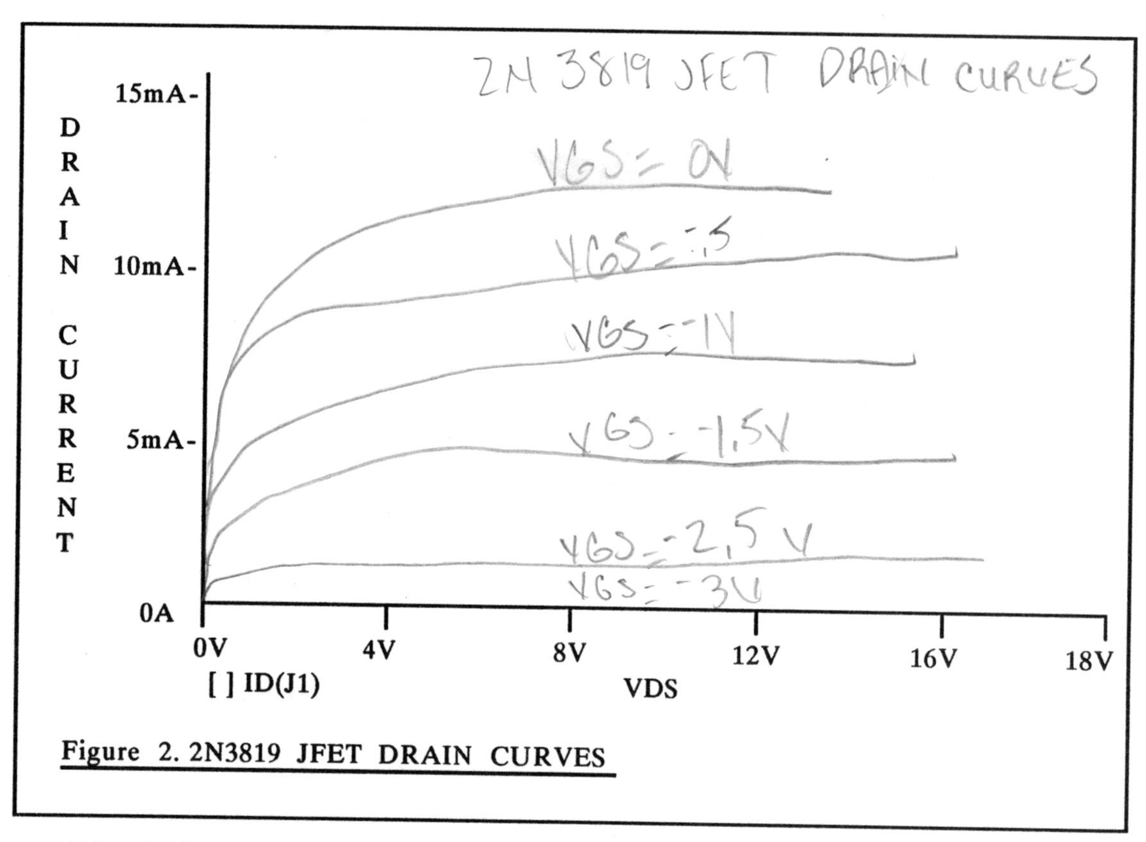

Figure 2. 2N3819 JFET DRAIN CURVES

Select Exit and Enter, and return to Control Shell.

(B) JFET Transconductance Curve

Transconductance in JFETs is a measure of the change in input gate voltage, VGS, versus the change in output drain current, ID. A transconductance parabolic curve is a plot of the drain current equation (p.9-11), and is presented in this exercise. A higher transconductance defines a higher change in output current for a given change in gate voltage. This can be measured on the transconductance curve at lower negative gate voltages in an N-channel JFET. As gate voltage is increased to pinchoff, less drain current is available, transconductance is lowered and greater non-linear operation is introduced.

Self-bias, one method of biasing a JFET, does not have an exact equivalent in the BJT. A source voltage is developed across the source self-bias resistor and is calculated by multiplying drain current by the source resistor. This self-bias voltage is negative feedback to the gate, measuring gate voltage in reference to the source, and sets the Q-point on the transconductance curve. Plots of several values of source resistance and corresponding gate-source voltages and drain currents will be presented with the Probe Graphical Waveform Analyzer program.

The Circuit File: Filename: JFETTCC

Enter the following netlist at the Circuit Editor.

```
*JFET TRANSCONDUCTANCE CURVES, J2N3819
VDS  2 0 DC 12V
VGS  1 0 DC -3.5V
J1  2 1 0 J2N3819
.DC VGS 0 -3.5 .1
.Model J2N3819 NJF(Vto=-3 Is=33.57f Beta=1.304m)
.LIB EVAL.LIB
.Probe
.End
```

Press Esc, to exit the Circuit Editor and save the file.

Netlist Notes

.DC VGS -0 -3.5 .1

In this DC sweep, gate input voltage will be stepped from 0Volts to -3.5Volts in .1Volt increments. Drain voltage, VDS, is constant at 12Volts.

.Model J2N3819 NJF(Vto=-3 Is=33.57f Beta=1.304m)

JFET parameters, Vto= -3, Is= 33.57f, and Beta= 1.304m, are copied from the library model for experience only. Adjusting these and all parameters, thereby writing a modified or new model, is accomplished simply by typing in the new data.

Run PSpice

Observe the monitor during DC analysis.

At Probe

Select Add_trace, and Enter.
Press F4 for a selection of circuit variables.
Arrow to ID(J1), and press Enter twice.
The JFET transconductance curve is displayed. From previous studies, the equation that plots the transconductance curve for all JFETs is,

$$ID = IDSS\ (1 - (VGS\ /\ VGS(off)))^2 \qquad \text{(Eq.1)}$$

for each value of gate voltage.

Note that higher transconductance, AC quantities id/vgs, in the equation, and later on the graph, occurs at lower gate voltages.

Set Ranges

The Y_axis range must be "set" to hold the drain current values.
Select Y_axis, and Enter.
Select Set_range, and Enter.
At Enter a Range: type 0mA 12mA , and Enter.
While at the Y_axis menu, place a title on the Y_axis.
Select Change_title, and Enter.
At Change_title: type Drain Current, and Enter.
Drain Current should now be the title along the left side of the graph.
Select Exit, and Enter, to leave the Y_axis menu.
Set the X_axis range: -3V 0V
Select Exit, and Enter, to leave the X_axis menu.

Self-Bias Source Resistance

The voltage drop across the source resistor in JFET circuits provides a gate-to-source negative feedback voltage differential. If the gate is tied to ground through a large resistor and held at 0V, the self-bias source voltage will set the required reverse bias necessary in JFETs and also set the circuit Q-point. This self-bias voltage is calculated,

V(GS) = -(ID)(Rs)

In this exercise, maximum drain current, IDSS, is 11.8mA and VGS(off) is -3V. This establishes the range of gate voltages at 0V to -3V. The source resistor to use for JFET optimum operation is calculated -VGS(off)/IDSS, or 254 Ohms. (This is the same equation used in solving for JFET drain resistance in the ohmic region)

Sketch Transconductance Curve and Self-Bias Lines

Using Equation 1, IDSS= 11.8mA, and VGS(off)= -3V, sketch the transconductance curve in Figure 3, page 9-14. Draw the self-bias line for Rs= 254 Ohms from coordinates IDSS= 11.8mA, VGS= -3V to the origin of the graph. The Q-point, where the self-bias line intersects the transconductance curve, should be approximately, VGS= -1.15V, ID= 4.4mA.

Next, using equation VGS= -(IDSS)(Rs), plot another self-bias line for a 100 Ohm source resistor. (VGS= -(11.8mA)(100 Ohms)= -1.18V) Draw this line from IDSS= 11.8mA, VGS= -1.18V to the origin. The Q-point coordinates for this line are approximately VGS= -.7V, ID= 7.2mA.

Probe JFET Self-Bias Lines

Compare your transconductance curve and Q-point measurements with the following plots drawn by Probe. First, let's insert the 254 Ohm source resistor, draw the self-bias line, and mark the Q-point.

Select Add_trace, and Enter.

At Enter variables or expressions, type -VGS/254, and Enter.

The Q-point should be approximately; VGS= -1.15V, ID= 4.4mA.

For a higher transconductance, gm, repeat with a 100 Ohm source resistor.

Select Add_trace, and Enter.

Type -VGS/100, and Enter.

Q-point coordinates are approximately, VGS= -.7V, ID= 7mA.

Adjusting the size of the source resistor moves the Q-point up and down the transconductance curve. For another example, select a 300 Ohm source resistor.

Select Add_trace, and Enter.

Type -VGS/300 , and Enter. Mark Q-point coordinates; VGS= ____, ID=____.

For a lower gm, repeat with a source resistance of 800 Ohms. Mark Q-point coordinates; VGS= ____, ID=____.

The display is a dramatic demonstration of PSpice simulation and analysis capabilities.

Labels

Label the Probe graph.

Place "2N3819 JFET TRANSCONDUCTANCE CURVE" in the upper center of the display by following the next three steps:

Select Label at Probe's initial menu, and Enter.

Select Text, Enter, type the label, and Enter.

Position the label.

Label the source resistance self-bias plots: Rs= 254 Ohms, Rs= 300 Ohms, Rs= 800 Ohms, and Rs= 100 Ohms.

Exit Probe.

Conclusion of Chapter 9

EOC Problem

(1) Repeat the experiments in the chapter by selecting other test levels. For example, adjust the sweep values and vary the input voltage and/or current.

Note: In control testing, it is usually best to change a minimum number of values at one time for easier tracking of experiment results. Detailed documentation is necessary with each circuit or analysis adjustment.

See "The Scientific Method in PSpice Experimentation," (p.iv).

Sketch Transconductance Curve

Sketch the transconductance curve, source resistance self-bias plots, and label the graph in Figure 3.

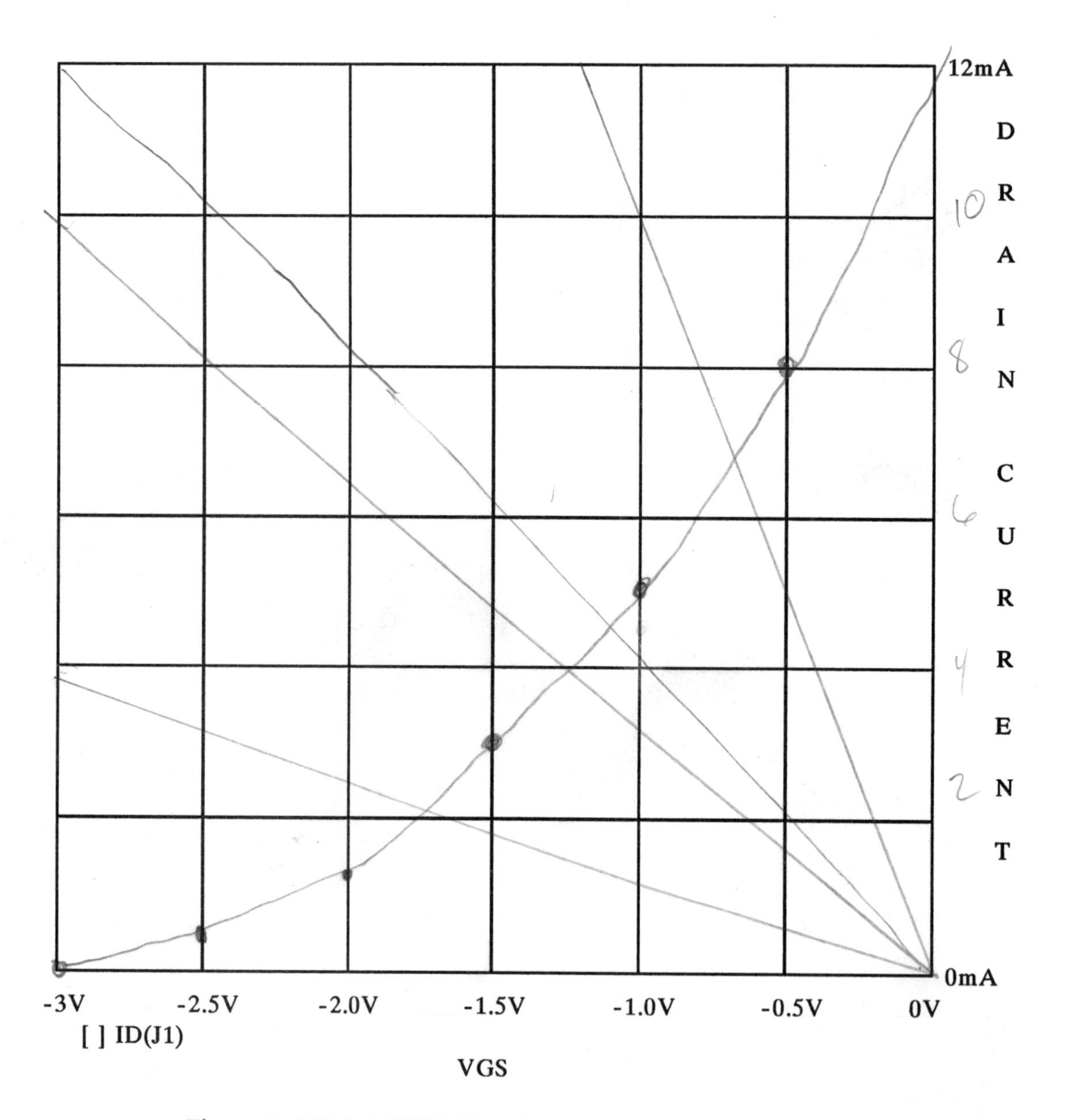

Figure 3. 2N3819 JFET TRANSCONDUCTANCE CURVE

Chapter 10

DIODE DOUBLERS, LIMITERS, DC CLAMPERS, PEAK-TO-PEAK DETECTORS, AND ZENER REGULATORS

Objectives: To use PSpice simulation/analysis and
Probe Waveform analysis evaluating:
(1) voltage doublers, (2) limiters,
(3) DC clampers, (4) peak-to-peak detectors, and
(5) Zener regulators

The diode, usually the first special device encountered in the study of electronics, is employed in a wide range of circuits. In this chapter, the doubler, limiter (or clipper), DC clamper, peak-to-peak detector, and zener regulator will be examined...each utilizing diode characteristics.

I. Full-Wave Voltage Doubler

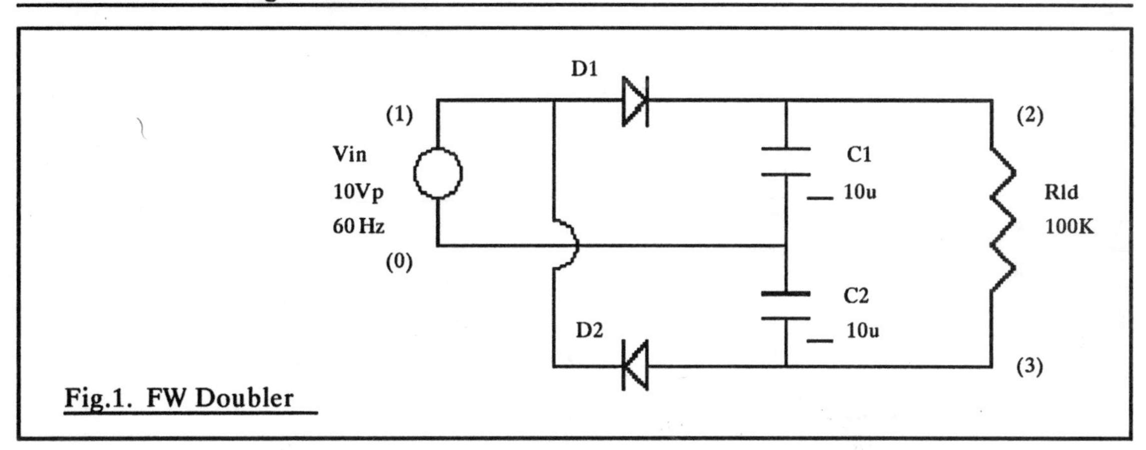

Fig.1. FW Doubler

Enter the following netlist at the Circuit Editor. Filename: FWVD

```
*FULL-WAVE VOLTAGE DOUBLER
Vin  1 0 SIN(0 10V 60)
D1   1 2 D1N4001
D2   3 1 D1N4001
C1   2 0 10u
C2   0 3 10u
Rld  2 3 100K
.Tran .34m 34.4m 0 .4m
.Model D1N4001 D(Is=10.0E-15 Rs=.1 Ikf=0 Cjo=1p N=1
+                Eg=1.11 Xti=3 Vj=.75 Fc=.5 Nr=2 Bv=100
+                Ibv=100.0E-6 Tt=5n Isr=100.0E-12)
.Probe
.End
```

Press Esc to save the circuit file.

Doubler diodes are rectifier, or power, diodes with ratings that exceed .5 watts. Small-signal diodes are rated at less than .5 watts and can be used in limiters, clampers, and peak-to-peak detectors. The 1N4001 diode model will be used in the experiments in this chapter. This diode model is not included in the evaluation library and must be written in the circuit netlist.

Run PSpice

Select Analysis, Enter, Run PSpice, and Enter.

At Probe

A disadvantage of the full-wave voltage doubler is the absence of a common input/output ground reference. The differential output is between V(2) and V(3).

Select Add_trace, and Enter.

Press F4, arrow to V(2), and Enter twice. Repeat for V(3).

Probe displays V(2) and V(3) at approximately +9.2V and -9.2V.

V(2) minus V(3) can be displayed to better observe voltage doubling. Select Remove_trace and clear the screen.

Select Add_trace, and Enter.

Type V(2) - V(3) at the bottom of the screen, and Enter.

The output is approximately 18.5 Volts.

Select Remove_trace and clear the screen.

Circuit performance is easily evaluated with all circuit waveforms displayed.

Select Add_trace, and display V(1), the input voltage, along with V(2) and V(3).

PSpice circuit analysis is impressively presented in these screens.

This concludes the full-wave voltage doubler exercise. Exit Probe.

II. Half-Wave Voltage Doubler

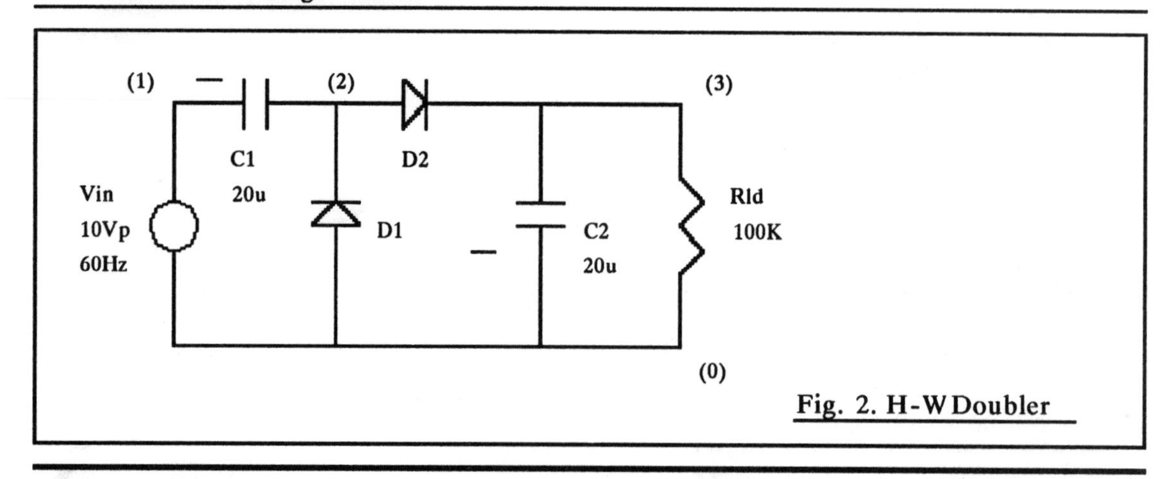

Fig. 2. H-W Doubler

Circuit filename: HWVD

The Netlist

```
*HALF-WAVE VOLTAGE DOUBLER
Vin  1  0  SIN(0  10V  60)
C1   2  1  20u            ;pos-neg
C2   3  0  20u            ;pos-neg
D1   0  2  D1N4001        ;anode-cathode
D2   2  3  D1N4001        ;anode-cathode
Rld  3  0  100K
.Tran  1.5m  128m  0  0
.Model  D1N4001  D(Is=10.0E-15 Rs=.1 Ikf=0 Cjo=1p N=1
+                  Eg=1.11 Xti=3 Vj=.75 Fc=.5 Nr=2 Bv=100
+                  Ibv=100.0E-6  Tt=5n  Isr=100.0E-12)
.Probe
.End
```

Press Esc to save the circuit file.

Netlist Notes

Transient analysis step_time and final_time are increased to allow C2 to completely charge. Step_time= 1.5ms, and final_time= 128ms.

Run PSpice

Select Analysis, Enter, Run PSpice, and Enter.

At Probe

Select Add_trace, and Enter.
Press F4, and arrow to V(3), the circuit output, and Enter.
C2 is fully charged after approximately 80ms and the output is approximately 18.5Volts. Study the display.
Select Add_trace, and display V(2) and V(1) along with V(3).
The half-wave voltage doubler circuit waveforms are displayed.
Again, PSpice analysis is impressive.
Exit Probe.

III. Limiters

A limiter, or clipper, clips a part of either the positive or negative portion of a waveform. If the positive part of the waveform is clipped, it is said to be a positive limiter or clipper.

Diodes used in these low current applications are called small-signal diodes. But we will continue to use the 1N4001 power diode because of its availability.

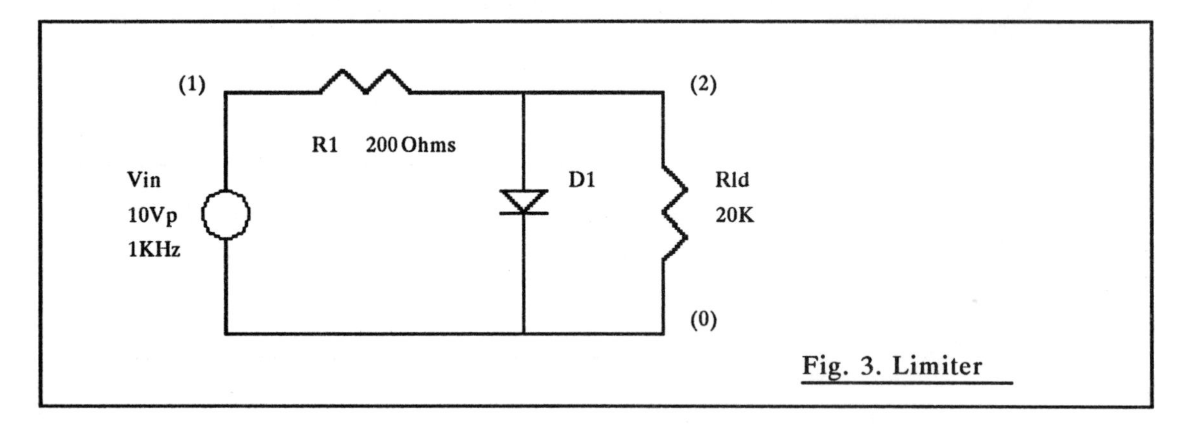

Fig. 3. Limiter

Filename: LIMITER

Netlist

```
*LIMITER
Vin   1  0  SIN(0  10V  1K)
R1    1  2  200Ohms
D1    2  0  D1N4001
Rld   2  0  20K
.Tran  .02m  2m  0  .01m
.Model  D1N4001  D(Is=10.0E-15 Rs=.1 Ikf=0 Cjo=1p N=1
+                Eg=1.11  Xti=3  Vj=.75  Fc=.5  Nr=2  Bv=100
+                Ibv=100.0E-6  Tt=5n  Isr=100.0E-12)
.Probe
.End
```

Press Esc to save the circuit file.

Netlist Notes

Transient analysis is set for 2 hertzs at 1KHz.
Diode current is expected to be less than 50mA peak.

Run PSpice

At Probe

Select Add_trace, and Enter.
Press F4, and arrow to V(2), the circuit output, and Enter twice.
Limiting of the waveform above .7Volts is noted.
Select Add_trace, and display the input, V(1), on the same screen.
Both input and output are displayed for circuit waveform comparison.
Study the outstanding Probe display.

Select Remove_trace, All, Exit, and Enter as needed.
Select Add_trace, and display I(D1).
The diode current waveform is measured at approximately 47mA peak.
Exit Probe

IV. Biased Limiter

The clipping level of the limiter can be adjusted by adding a bias voltage in series with the clipping diode. In this experiment, a sine wave will be clipped on both alternations to provide a near square wave output.

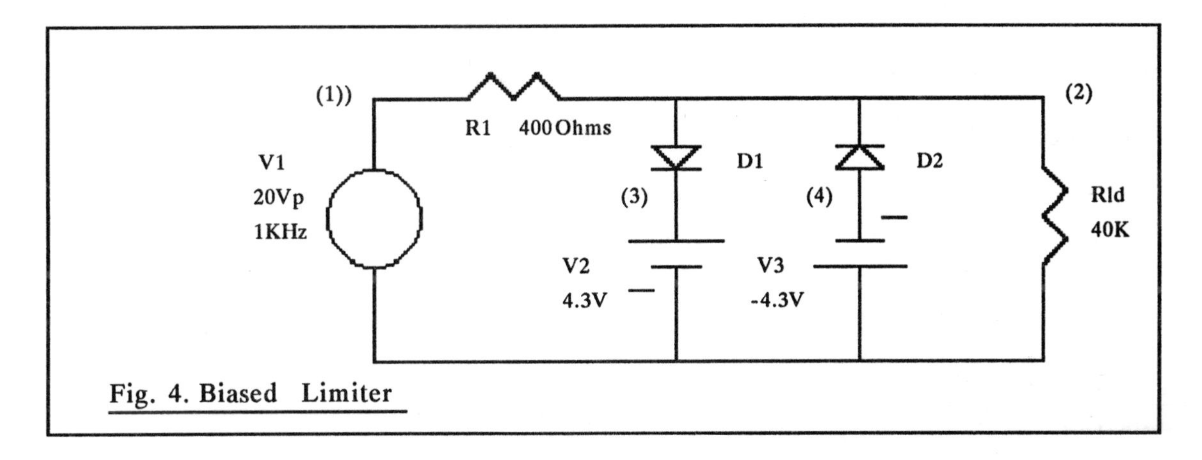

Fig. 4. Biased Limiter

Filename: BIASLMTR

The Netlist

```
*BIASED LIMITER
V1   1 0 SIN(0  20V  1K)
V2   3 0 DC  4.3V
V3   4 0 DC  -4.3V
R1   1 2 400Ohms
D1   2 3 D1N4001     ;anode-cathode
D2   4 2 D1N4001
Rld  2 0 40K
.Tran .02m  2m  0  .01m
.Model  D1N4001  D(Is=10.0E-15 Rs=.1 Ikf=0 Cjo=1p N=1
+                  Eg=1.11 Xti=3 Vj=.75 Fc=.5 Nr=2 Bv=100
+                  Ibv=100.0E-6  Tt=5n Isr=100.0E-12)
.Probe
.End
```

Press Esc to save the circuit file.

Netlist Notes

Diode anodes are connected to the first node in the statement.

Run PSpice

Select Analysis, Enter, Run PSpice, and Enter.

At Probe

Select Add_trace, and Enter.

Press F4, and arrow to V(2), the circuit output, and Enter twice. Limiting of the positive and negative alternations at Vsource + Vd is noted. (Vs=4.3V, Vd=.7V) The waveform amplitude is 5Vpeak.

Add_trace, and place V(1), the input stimulus, on the same screen. Also, add V(3) and V(4) to observe all circuit measurements. The display is an illustration of the Probe Graphical Waveform Analyzer's capabilities.

Remove all traces, and examine diode currents, I(D1) and I(D2). Diode current is limited to less than 40mA. The current waveform displays the portion of the incoming stimulus that was clipped.

Exit Probe.

V. DC Clamper or Restorer

A DC clamper adds a DC voltage to the input waveform. If the incoming signal swings around zero, a positive clamper will move the waveform above zero. Negative clampers move the waveform below the zero level. In this experiment, a 20Vpeak input signal will be clamped at near the 20V DC level at the output. Reversing diode node connections changes the polarity of the clamper.

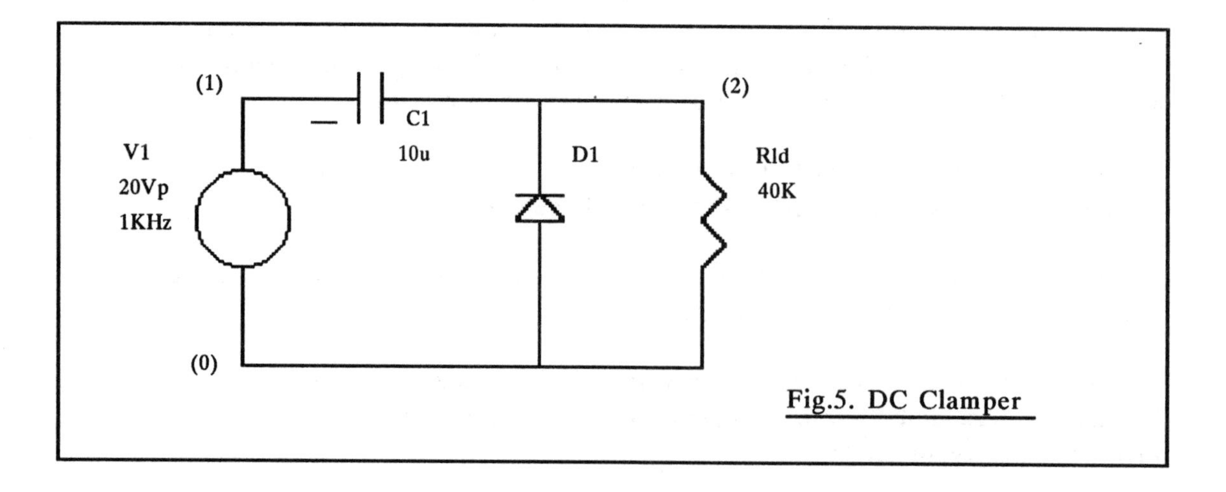

Fig.5. DC Clamper

Filename: DCCLAMP

The Netlist

```
*DC CLAMPER OR RESTORER
V1  1 0 SIN(0  20V  1K)
C1  1 2 10u
D1  0 2 D1N4001    ;anode-cathode
Rld  2 0 40K
.Tran  .02m  2m  0  .01m
.Model  D1N4001  D(Is=10.0E-15 Rs=.1 Ikf=0 Cjo=1p N=1
+                  Eg=1.11 Xti=3 Vj=.75 Fc=.5 Nr=2 Bv=100
+                  Ibv=100.0E-6  Tt=5n  Isr=100.0E-12)
.Probe
.End
```

Netlist Note

Initial diode surge current is above one ampere for a short period of time.

Run PSpice

Select Analysis, Enter, Run PSpice, and Enter.

At Probe

Select Add_trace, and Enter.
Press F4, and arrow to V(2), the circuit output, and Enter twice.

Probe displays the output waveform clamped above zero volts.

Select Add_trace, and display the input waveform, V(1), on the same screen. As the input swings around zero, the charge of C1 builds to effectively clamp the output at near the 20V DC level. The peak-to-peak swing is approximately 40V to near 0V. A clear illustration of a DC restorer or clamper.

Select Remove_trace, and clear the screen.
Select Add_trace, and display I(D1), the diode current.

Initial input current of 1.25A charges C1, which remains charged, and diode current levels off for circuit operation.

Return to the Circuit Editor and reverse the diode node connections. Repeat the same exercise and observe the operation of a negative clamper. Study the waveforms in support of theory.

IV. Peak-To-Peak Detector

A DC clamper and a rectifier combine to form a peak-to-peak detector. The circuit first shifts the input signal to a level above (or below) zero. The next part of the circuit is a rectifier. DC output is approximately double the peak input amplitude.

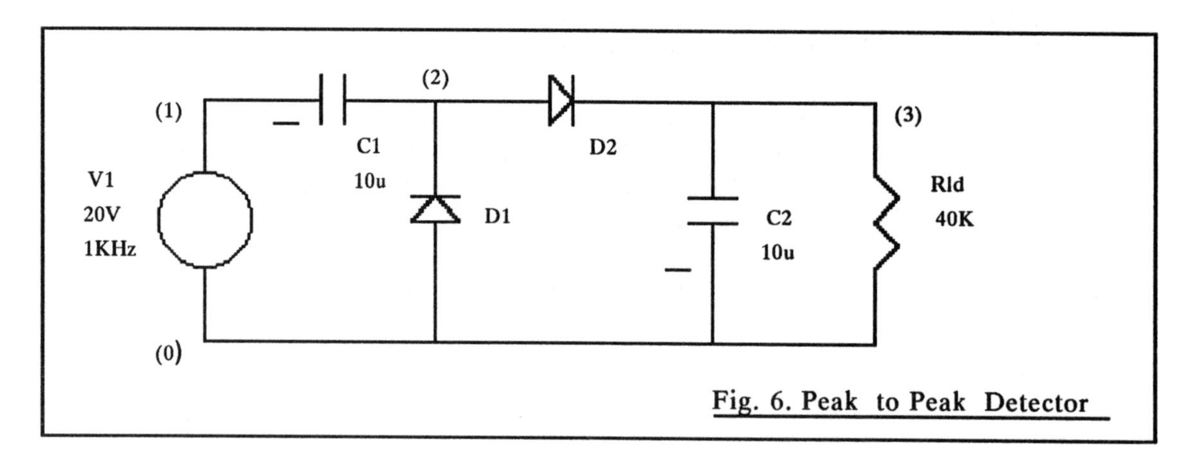

Fig. 6. Peak to Peak Detector

Filename: PPDETECT

Netlist

```
*PEAK TO PEAK DETECTOR
V1   1  0  SIN(0  20V  1K)
C1   2  1  10u
D1   0  2  D1N4001     ;anode-cathode
D2   2  3  D1N4001
C2   3  0  10u
Rld  3  0  40K
.Tran  .08m  10m  0  0
.Model  D1N4001  D(Is=10.0E-15 Rs=.1 Ikf=0 Cjo=1p N=1
+                  Eg=1.11  Xti=3  Vj=.75  Fc=.5  Nr=2  Bv=100
+                  Ibv=100.0E-6  Tt=5n  Isr=100.0E-12)
.Probe
.End
```

Netlist Notes

Transient analysis final_time has been extended to 10 cycles of the input stimulus. Initial charge of the capacitors takes approximately 8ms. Print_step time has been set to .08ms.

Initial diode surge current is above one ampere for short periods of time.

Run PSpice

Select Analysis, Enter, Run PSpice, and Enter.

At Probe

Select Add_trace, and Enter.
Press F4, and arrow to V(3), the circuit output, and Enter twice.
The output increases to above 38V by the 8ms point in the analysis. Capacitor step charge is evident in the first 7ms.
Select Add-trace, and display the input stimulus, V(1).
Select Add-trace, and display DC clamper action at V(2).

The Probe Waveform Analyzer display is characteristic of the tremendous worth of computer aided circuit design. Each section of the circuit's operation can be evaluated and improved by examination of this simulated analysis.
Select Remove_trace, and clear the screen.
Select Add_trace, and display diode currents I(D1) and I(D2).
These initial current spikes should be considered when selecting small-signal diodes.
Select Remove_trace, and clear the screen.
Display capacitor currents, I(C1) and I(C2).
Select Exit, and Enter.
Exit Probe.

VII. Zener Regulator

The zener diode used in this exercise is held in the breakdown region to assure output voltage regulation. .DC sweep is used to simulate an unregulated input.

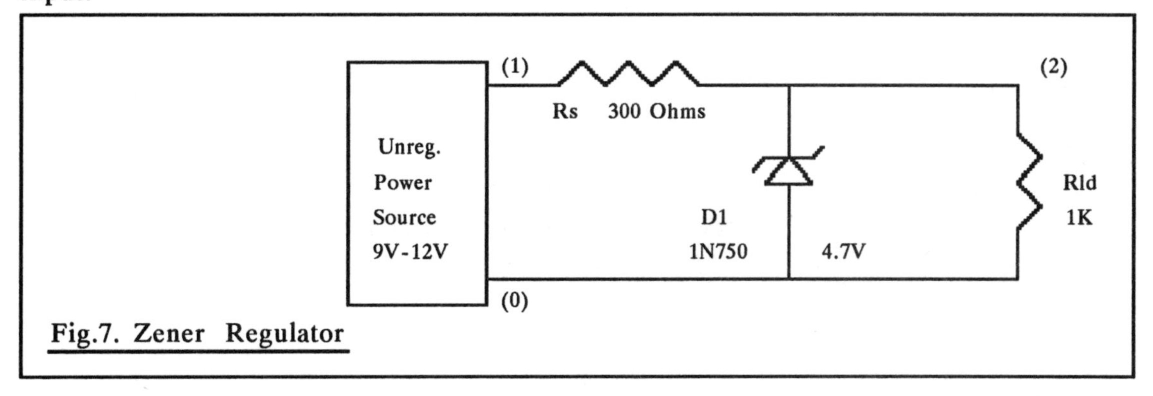

Fig.7. Zener Regulator

Netlist; Filename: ZENERREG

```
*ZENER REGULATOR
Vin  1  0  DC  10V
Rs   1  2  300Ohms
D1   0  2  D1N750     ;Zener, 4.7V
Rld  2  0  1K
.Model  D1N750  D(Is=880.5E-18 Bv=4.7 Ibv=20m)
.DC  Vin  9  12  .2    ;sweep  input;  9V - 12V, in .2V steps
```

Cont:

Cont.

```
.LIB EVAL.LIB
.Print DC V(1) V(2)
.Print DC I(Rs) I(Rld) I(D1)
.Probe
.End
```

Press Esc to exit Circuit Editor and save the circuit file.

Netlist Notes

.Model D1N750

Zener model included in the evaluation library. Vz= 4.7V.

.DC Vin 9 12 .2

This control statement will sweep the input to the zener regulator, 9V to 12V in .2V steps. A large swing is used for illustration purposes.

Estimated Maximum Currents (Vin= 12V)

I(Rs)= I(Vin)= (Vin-Vz)/Rs	I(Rld)= V(Rld)/Rld	Iz= I(Vin)-I(Rld)
= (12V-4.7V)/300	= 4.7V/1K	= 24.3mA-4.7mA
= 24.3mA	= 4.7mA	= 19.6mA

Estimated Minimum Currents (Vin= 9V)

I(Rs)= I(Vin)= 14.3mA	I(Rld)= 4.7mA	Iz= 9.6mA

Run PSpice

At Probe

Examine V(1) for input sweep, and V(2) for zener control.

Select Remove_trace, and clear the screen.

Examine I(Rs), I(Rld), and I(D1), separately, for minimum and maximum values and compare with calculations. Exit Probe.

Examine the Output file tables for these values.

Exit Output file.

Conclusion of Chapter 10

EOC Problems

(1) Experiment with the circuits in this chapter using PSpice and Probe by:
(a) reversing the diodes, (b) adjusting capacitor size,
(c) changing load value, Rld, or (d) changing input voltage.
Complete calculations to predict changes before PSpice runs.

(2) Design a circuit that uses an LED to indicate a power-on condition. Limit LED current to 8mA. Circuit load current is 1.8A. Supply all component values for the design. Complete all voltage, current, and power estimates, then use PSpice to evaluate the circuit. Vs= 20V DC. (Use D1N4001 model to simulate LED)

Chapter 11

BIPOLAR JUNCTION TRANSISTOR

DC BIAS AND CLASS A OPERATION

Objectives: (1) To demonstrate PSpice simulation and bias analysis of bipolar junction transistor circuits
(2) To use Output file data to check Q-point bias settings

Previous studies have shown the necessity for using DC voltage when biasing bipolar junction transistors into operation. A voltage divider will be used in this exercise for bias settings of a typical Class A amplifier. (Fig.1) Saturation and cutoff values will be calculated and a DC load line will be drawn for the circuit. Next, the quiescent point, or Q-point, will be located on the load line. The Q-point will then be reset for optimum operation on the DC load line. (AC load line settings will be considered in the next chapter) Since Q-point location is circuit dependent, rather than device dependent, it can be adjusted by varying component values and/or circuit voltages. PSpice will simulate and analyze the circuit as we progress and record measurements in the Output file. This data will be compared with circuit calculations.

Circuit Expectations

The process of common-emitter amplifier design begins by (1) defining circuit objectives, (2) choosing a suitable transistor, and (3) calculating bias values. In this exercise, we will analyze, rather than design, the CE circuit on the next page for an expected gain of 5 to 10. We will use the Q2N2222A NPN library model with the Q1 4 2 3 Q2N2222A circuit statement. (4-collector, 2-base, 3-emitter) Only three of more than fifty parameters will be defined to set up the model. They are ideal maximum forward beta (bf=100), base-collector p-n capacitance (cjc=10p), and maximum base resistance (rb=20 ohms). The remaining default values are accepted by not assigning values to them in the model statement. A list of parameter definitions is included in Appendix B. The device library listing for the Q2N2222A model is recorded below. (.Model Q2N2222A NPN is not an exact model in the evaluation version)

```
.Model  Q2N2222A  NPN(Is=14.34f  Xti=3  Eq=1.11  Vaf=74.03
+                     Bf=255.9  Ne=1.307  Ise=14.34f  Ikf=.2847
+                     Xtb=1.5  Br=6.092  Nc=2  Isc=0  Ikr=0  Rc=1
+                     Cjc=7.306p  Mjc=.3416  Vjc=.75  Fc=.5
+                     Cje=22.01p  Mje=.377  Vje=.75  Tr=46.91n
+                     Tf=411.1p  Itf=.6  Vtf=1.7  Xtf=3  Rb=10)
```

The Test Circuit

A sketch of the circuit is presented below. It will assist in defining nodes and visualizing design changes. Next, all circuit bias calculations will be evaluated to compare with PSpice simulation and analysis later.

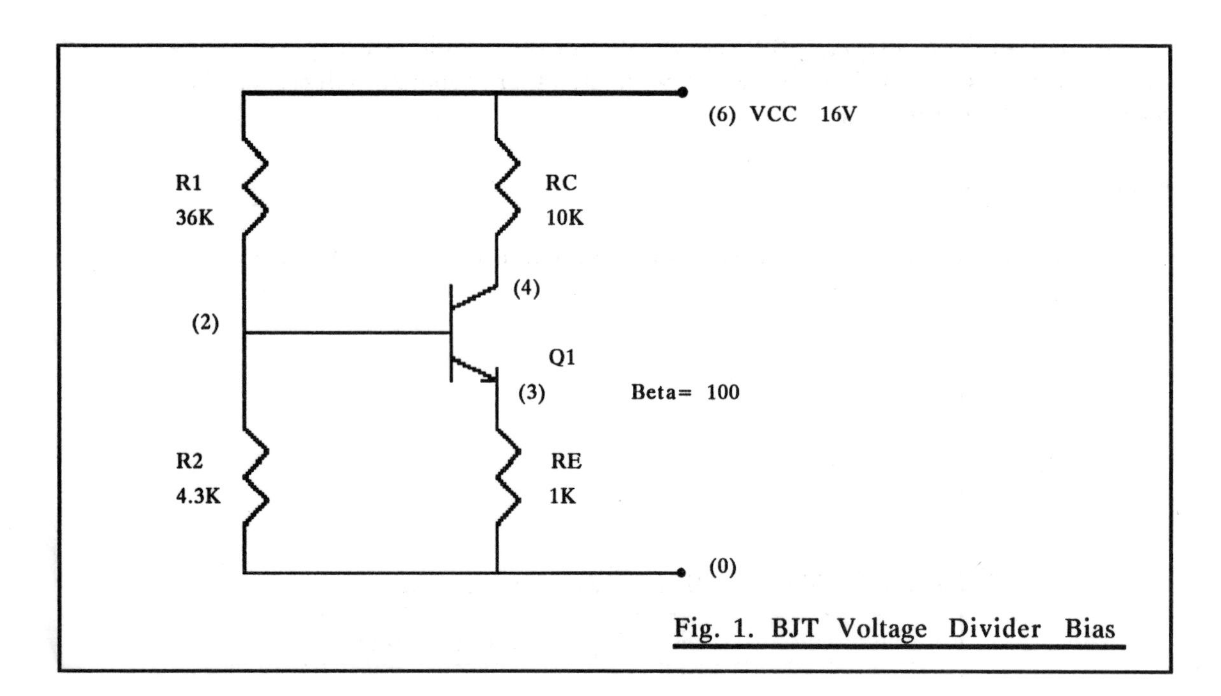

Fig. 1. BJT Voltage Divider Bias

Bias Calculations

First, compute all circuit voltages and currents and list these values in Table I.

Node 2 base voltage is set at 1.7Volts by the voltage divider network, R1 and R2. Emitter voltage, Node 3, is estimated at .7V less than VB at 1 Volts. Calculate IE and IB.

IE = VE / RE = 1V / 1K = 1 mA (Initial Q-point current)
IB = IE / DC Beta = 1mA / 100 = 10uA

IC is estimated to approximate IE. Calculate collector voltage and collector-emitter voltage.

VC = VCC - (IC x RC) = 6V
VCE = VC - VE = 5V (Initial Q-point value)

Voltage Divider Current, I(R2)

Voltage divider current should be at least 20 times the base current to insure a stable base voltage.

Minimum Ivd > 20 x IB = 20 x .01mA = .2 mA

Actual Ivd = VCC / (R1 + R2) = 16V / 40.3K = .4 mA

Voltage divider current, I(R2), is included in the circuit netlist .print statement on page 5. This value, along with the other requested values, will be recorded in the Output file.

Power Dissipation

Voltage source current and total power dissipation is also calculated by PSpice. Our circuit estimates are:

I(VCC) = Ivd + IC
= .4mA + 1mA = 1.4mA

P(VCC) = 16V x 1.4mA
= 22.4mW

Table I. Calculated and PSpice Output File Values

Circuit	Calculated	PSpice Output Values
VCC	16 V	16 V
I(VCC)	1.4 mA	1.28 mA
P(VCC)	22.4 mW	20.48 mW
V(2)	1.7 V	1.673 V
V(3)	1 V	.901 V
V(4)	6 V	7.075 V
I(R2),(Ivd)	.4 mA	.389 mA
I(RE)	1 mA	.901 mA
I(RC)	1 mA	.892 mA
I(B)	10 uA	9 uA (IRE - IRC)

DC Load Line: Saturation Current And Cutoff Voltage

At saturation, the voltage across the collector resistor, RC, is equal to the power supply voltage minus collector voltage which is at emitter potential.

Calculate the voltage across RC and collector saturation current.

VRC = VCC - VE = 16V - 1V = 15V

IC(sat) = (VCC - VE) / RC = (16V - 1V) / 10K = 1.5mA

At cutoff, the collector is at supply voltage potential, 16V, and the collector-emitter voltage defines the lower end of the DC load line.

VCE(cutoff) = VCC - VE = 16V - 1V = 15 Volts

Load Line Graph

A graph of the DC load line is drawn to visually observe setting the Q-point. Sketch the load line and point the labels.

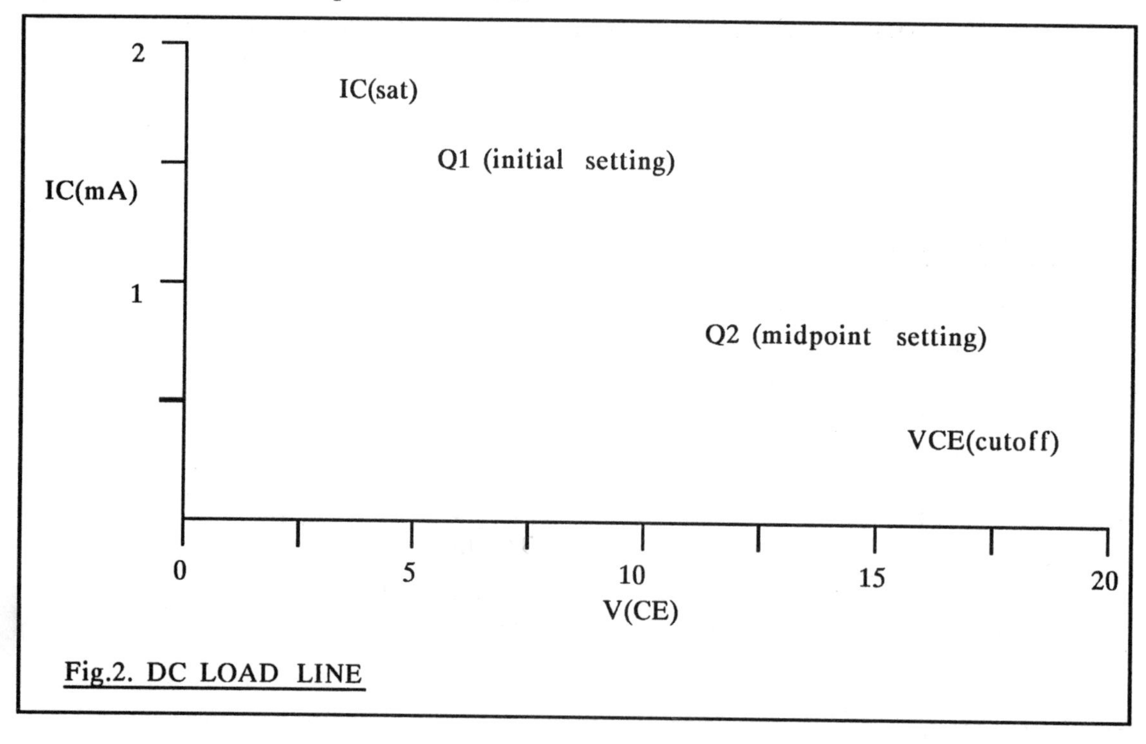

Fig.2. DC LOAD LINE

PSpice "Breadboarding" and Analysis

At PSEVAL52 directory, type PS, and Enter.

At Control Shell, Select Files, Enter, Current File, and Enter.
Filename: BJTBIAS1

The circuit file (netlist) is the PSpice "breadboard." Simulation and analysis by PSpice is performed on this circuit (.CIR) file.

The Netlist

```
*BJTBIAS1 - VOLTAGE DIVIDER BIAS CIRCUIT
VCC  6 0 DC 16V
R1   6 2 36K
R2   2 0 4.3K
Q1   4 2 3 Q2N2222A
RC   6 4 10K
RE   3 0 1K
.MODEL Q2N2222A NPN(bf=100 cjc=10p rb=20)
.LIB EVAL.LIB      ;search the EVAL.LIB for .MODEL Q2N2222A NPN
.DC VCC 16 16 1 ;DC sweep begins and ends at 16V, in 1V data steps
*The following statement programs PSpice to print values listed.
*(PSpice would automatically print all node bias voltages
* if a .print statement was not included)
.Print DC V(2) V(3) V(4)
.Print DC I(R2) I(RC) I(RE)  ;PSpice will print these values
*                              from DC sweep analysis.
*                  End of the circuit file. Another file could
.End             ;begin immediately after this .END statement.
```

Save and Run PSpice

Press Esc, and exit the Circuit Editor.
Correct errors as prompted by PSpice.

Select Analysis, and Enter.
Select Run PSpice, and Enter.

Note: PSpice will signal that Probe is abandoned and will return to the Control Shell. Everything is A-OK. The Control Shell "Run PSpice" program defaults to a Probe run, but Probe needs variables to operate and the sweep specified is a constant 16Volts.

Browse the Output File

Select Files, Browse Output, and Enter.
PgDn, PgUp, and the Arrow keys move within the Output file.

Output File paging:

(1) BJTBIAS1 CIRCUIT DESCRIPTION
(2) BJT MODEL PARAMETERS
(3) DC TRANSFER CURVES, TEMP (Voltage values)
(4) DC TRANSFER CURVES, TEMP (Current values)

Record the following Output file data here and in the PSpice Output Values column of Table I.

V(2) = ______V V(3) = ______V V(4) = ______V

I(R2)= ______mA I(RC)=______mA I(RE)= ______mA

The .DC VCC 16 16 1 sweep statement must be disabled for the next two measurements. Make the line a comment statement by inserting an asterisk.

Run PSpice and record the following values here and in Table I. Notice that all node voltages are printed. Remove the asterisk when measurements have been recorded.

Voltage Source Current = ______mA

Total Power Dissipation = ______mW

Are these readings in alignment with calculated DC bias values in Table I?

Q-Point Adjustment

For Class A operation, the Q-point should be moved to the center of the load line for optimum operation. To accomplish this move we estimate the center of the DC load line to be one-half IC(sat).

New Q-point setting = IC(sat) / 2
= 1.5mA / 2 = .75mA

The Q-point is defined by this new collector current, .75mA. Resetting the collector current can be achieved by changing the values of the voltage divider network, adjusting VCC, or more easily, by changing the size of the emitter resistor.

RE = 1V / .75mA
= 1.33 KOhms

New DC Bias Calculations

Base voltage (Node 2) is set by the voltage divider network, R1 and R2, at 1.7 Volts. Emitter voltage (Node 3) remains at 1 Volts.

The following values are recorded in Table II.

IE = VE / RE = 1V /1.33K = .75mA

IB = IE / DC Beta = .75mA / 100 = 7.5uA

IC is estimated to approximate IE.

VC = VCC - (IC x RC) = 8.5V
VCE = VC - VE = 7.5V

I(VCC) = Ivd + IC
= .4mA + .75mA = 1.15mA

P(VCC) = 16V x 1.15mA
= 18.4 mW

Table II. Calculated and PSpice Output File Values

Circuit	Calculated	PSpice Output Values
VCC	16 V	______V
I(VCC)	1.15 mA	______mA
P(VCC)	18.4 mW	______mW
V(2)	1.7 V	______V
V(3)	1 V	______V
V(4)	8.5 V	______V
I(R2) (Ivd)	.4 mA	______mA
I(RE)	.75 mA	______mA
I(RC)	.75 mA	______mA
I(B)	7.5 uA	______uA (IRE - IRC)

Return to the Circuit Editor and change the value of RE to 1.33K.

Run PSpice

Browse the Output file and record the following data:

V(2) = ______V V(3) = ______V V(4) = ______V

I(R2)= ______mA I(RC)= ______mA I(RE)= ______mA

I(VCC)= ______mA

Disconnect .DC sweep, Run PSpice and record:

I(VCC)= ______mA P(VCC)= ______mW.

Enter these values in Table II.
Is this data in alignment with the new bias calculations?

Reset Q-point on the Graph

Enter the new Q-point on the graph. Draw a perpendicular line from the operating point down to the X_axis and a horizontal line over to the Y_axis. Is the Q-point near the center of the DC load line? The new settings should bias the transistor for Class A operation.

In the next chapter, BJT AC LOAD LINE, PSpice will be used to evaluate how close to the midpoint the Q-point has been set.

Conclusion of Chapter 11

EOC Problems

(1) Experiment with the circuit by moving the Q-point up and down the DC load line. Change RE.

(2) Repeat Problem 1 by changing R1.

(3) Repeat Problem 1 by changing the value of VCC.

Note: Complete all calculations for each change before running PSpice analysis.

Chapter 12

BJT AC LOAD LINE

Objectives: (1) To check BJT Q-point mid-range operation using PSpice
(2) To calculate, then simulate AC load line considerations
(3) To measure voltage gain and dB gain
(4) To practice using PSpice programs

In the previous chapter, the Q-point of a bipolar junction transistor common-emitter amplifier was set to mid-range on the DC load line. With the Q-point at mid-point, maximum collector swing along the DC load line is allowed before output waveform clipping occurs. In this chapter, using PSpice simulation and analysis, we will check this mid-range setting and then examine AC load line operation. We will use the BJTBIAS1 circuit from Chapter 11.

Type PS at the PSEVAL52 directory, and Enter.

Select Files, Enter, Current File, and Enter.
Type: BJTBIAS1 , and Enter.
Select Files, Edit, and Enter.
Edit the netlist as listed.

The Netlist

```
*BJTBIAS1 - VOLTAGE DIVIDER BIAS CIRCUIT
VCC  6  0  DC  16V
R1   6  2  36K
R2   2  0  4.3K
Q1  4  2  3  Q2N2222A
RC  6  4  10K
RE  3  0  1.33K
.Model  Q2N2222A  NPN(bf=100  cjc=10p  rb=20)
.LIB  EVAL.LIB
.DC  VCC  16  16  1
.Print  DC  V(2)  V(3)  V(4)
.Print  DC  I(R2)  I(RC)  I(RE)
.End
```

Study the netlist in reference to the circuit drawing on page 12-2.

We will return to the Circuit Editor later to edit the input stimulus and to insert the output load.

Press Esc, and save the circuit file.

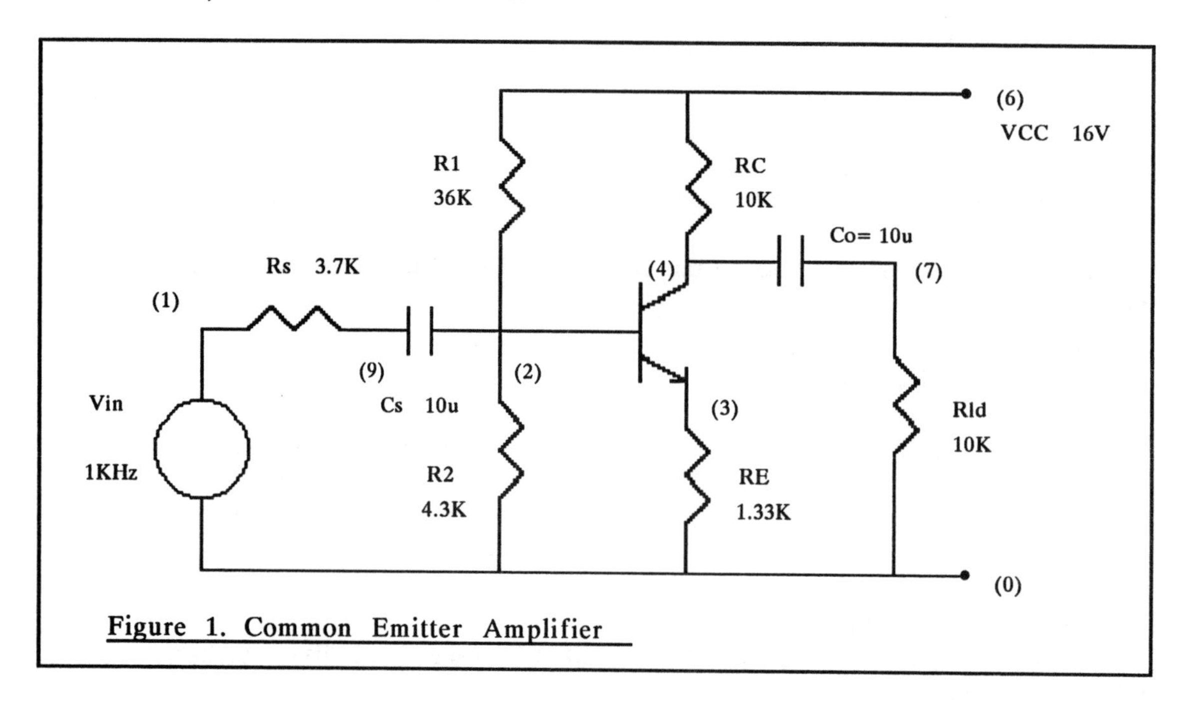

Figure 1. Common Emitter Amplifier

A graph of the DC load line is drawn to aid in constructing the AC load line.

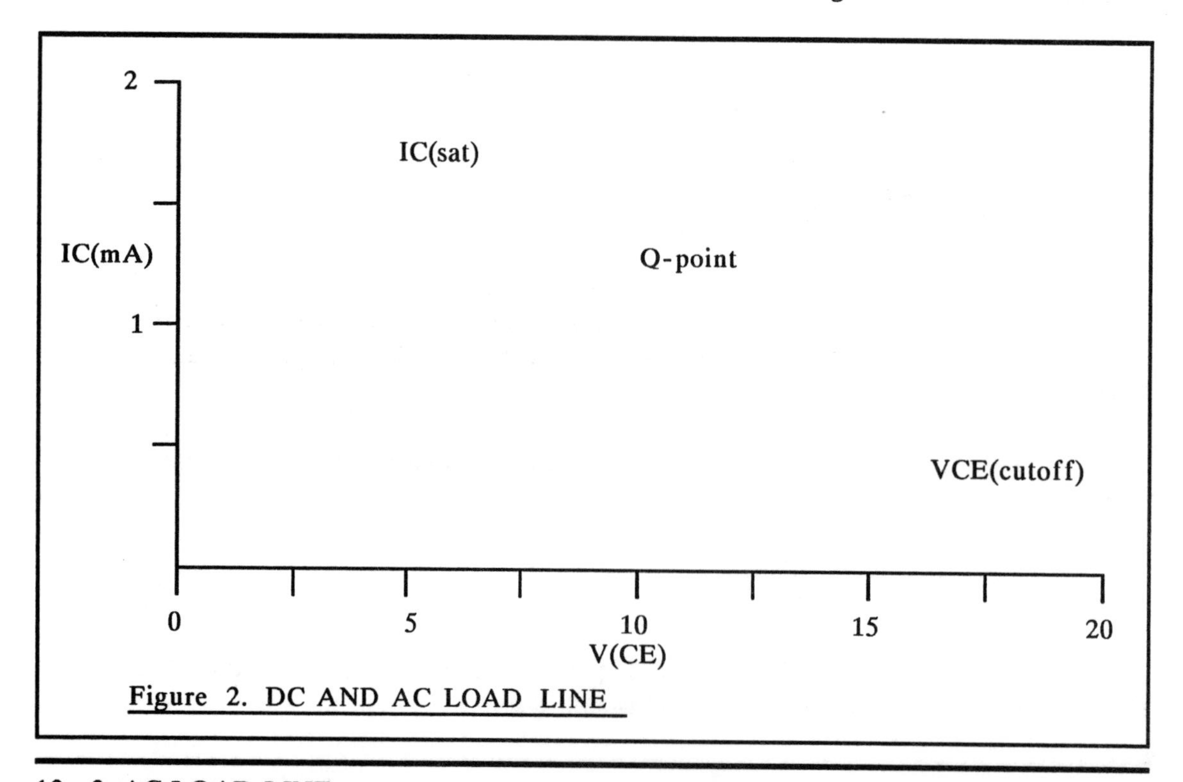

Figure 2. DC AND AC LOAD LINE

Circuit data from the previous chapter included:

VCE(cutoff) = VCC - VE
= 15V

IC(sat) = (VCC - VE) / RC
= 1.5mA

IC(Q-point) = IE = .75mA
VCE(Q-point) = VC - VE = 7.5V

Draw the DC load line and set the Q-point. The AC load line will be added later.

DC Load Line Estimated Gain

With RE=1.33K and RC=10K, a gain of 7.5 is expected (RC/RE). Gain expressed in decibels is 17.5 dB (dB gain = 20 log (7.5) = 17.5). An input amplitude of 1Vpeak at the base should be just enough to start clipping the output waveform at saturation and cutoff. If the amplifier is biased at the midpoint of the DC load line, clipping should be approximately equal on both swings of collector voltage.

Next, an input stimulus, Vin, will be edited into the circuit file with its related components.

Adding An Input Stimulus

Using the Circuit Editor, add the following stimulus to the circuit file. Insert the lines just after the VCC entry.

```
Vin  1  0  AC  1  SIN(0  2V  1KHz)
Rs   1  9  3.7K
Cs   9  2  10u
```

The input impedance of the amplifier is approximately the same as the source resistance (3.7K). Therefore, setting Vin to 2Vpeak will deliver 1Vpeak to the base.

Add the transient analysis and AC sweep control statements after the library line.

```
.Tran  .02ms  2ms  0  .01ms
.AC  DEC  3  50m  10Meg
```

The transient analysis line programs PSpice to examine two hertz of the 1KHz signal. (One hertz at 1KHz= 1ms) Final time= 2ms.

The .AC sweep statement will provide a small-signal analysis of the amplifier. A decade sweep is stated with 3 points per decade, beginning at .05Hz and ending at 10MHz. AC 1 sets the input level for this analysis.

A PSpice Note: PSpice permits adding transient analysis and small-signal analysis control statements anywhere in the circuit file...between the title line and the .End statement. Include the (.) at the beginning of the line.

Transient Analysis Gain Measurement

Run PSpice

Select Analysis, Enter, Run PSpice, and Enter. The Probe title screen should be displayed when the analyses are completed. If not, check the circuit file.

At Probe Title Screen:
Select Transient_analysis, and Enter.
Select Add_trace, and Enter.
Press F4 for a list of variables and expressions available.
Study the various PSpice measurements that are available for Probe presentation.
Arrow to V(4), collector output, and Enter twice.

The ac collector voltage is displayed as it swings toward saturation and cutoff. A small distortion of the sine wave is evident at amplitude extremes. The upper alternation, approaching cutoff, is driven to 16V, the limits of the power supply. The lower alternation is driven toward saturation, less than 2V. DC offset voltage can also be measured at near 9V.

Record vc=____Vp. (Should be approximately 7Vp)

If Vin is increased, clipping would, of course, be even more evident. We will edit Vin to 2.5Vp later in the exercise. The base input will be 1.25Vp and the output waveform will reflect saturation.

Voltage Gain

Next, check the input at the base, V(2), and calculate gain.
Select Remove_trace, and clear the screen.
Select Add_trace, and Enter.
Press F(4), and arrow to V(2), and Enter twice.
The screen should display a base signal voltage of approximately 1Vpeak.

Record vb= ____Vp. Note the offset voltage, VB= 1.7V.

Calculate vc/vb. Gain= ____. (Approximately 7)
Select Exits, and Enter, to return to Control Shell.

AC_Sweep Measurements

Voltage Gain and Decibel Gain

Return to the Circuit Editor and edit Rs to 2 Ohms, for the following measurements. AC_sweep input is defined by the AC 1 listing, or unity input, in the Vin statement. This will place an input signal of 1V amplitude on the base of the transistor. The ratio of base voltage to collector voltage will be expressed as voltage gain and dB gain in AC_sweep analysis.

Edit the following line with the Circuit Editor:

```
Rs   1  9  2Ohms
```

Press Esc to save the file.

Run PSpice

Voltage Gain Measurement

At Probe Title Screen:

Select AC_sweep, and Enter.

Note: If DC_sweep is selected by mistake, PSpice will exit Probe for lack of sweep data. To return, select Probe at Control Shell, and Enter.

At Probe Menu:

Select Add_trace, press F4, arrow to V(4), the collector output, and Enter twice.

A gain of approximately 7 is measured by AC_sweep small-signal analysis. The sweep response ranges from ~10Hz to above 1MHz. Decibel gain is the next measurement and the equation for dB gain is...dB gain = 20 log (7) = 16.9.

Select Remove_trace, and clear the screen.

dB Gain Measurement

Select Add_trace, and Enter.

Type Vdb(4), and Enter. The trace indicates the expected gain at approximately 17 dB over the sweep range. Reset Y_axis range for a better presentation.

Select Y_axis, Enter, Set_range, and Enter.

At Enter a range: type 0 20 , and Enter.

Exit Probe after recording dB gain. Gain= ____dB.

Edit Rs to the original setting:

Rs 1 9 3.7K

Press Esc to save the edited file.

The Stimulus Editor will be used in the next section to edit Vin from 2Vpeak to 2.5Vpeak for the purpose of further overdriving the amplifier. The Q-point midpoint setting will be made more apparent by the equally clipped alter-nations.

Use StmEd to Edit Vin (To overdrive the amplifier)

Select StmEd at Control Shell, and Enter.
Select Edit, and Enter.
Accept "S", and Enter, to save all changes.

The Stimulus Editor screen displays Vin at 2Vpeak.
Select Modify_stimulus, and Enter.
Select Transient_parameters, and Enter.

Step through each of the parameters, leaving each one unchanged except 2)VAMPL. At 2)VAMPL, edit the amplitude of the input stimulus.
Backspace and erase 2, and type 2.5, and Enter.
Select Exit, and Enter.

The new stimulus is displayed at 2.5Vpeak.

Select Exit, and Enter, to leave StmEd edit screen.
Select Exit, and Enter, to leave StmEd menu screen.
Select Exit_program, and Enter, to leave StmEd.

Note: StmEd writes this modified stimulus into the circuit file. Take a look at the edited file with the Circuit Editor. Select Files, Edit, and Enter.
Press Esc to return to Control Shell.

Next, we will Run PSpice to include the new stimulus setting at 2.5Vp.

Run PSpice (To observe increased distortion)

At Probe Title Screen:

Select Transient_analysis, and Enter.

At Probe Menu:

Select Add_trace, Enter, and press F4.

Arrow to V(4), the collector output, and Enter twice.

Clipping of the output waveform is quite evident as the input signal overdrives the amplifier. The Q-point is located near the mid-point of the DC load line as indicated by equal clipping of the waveform. Note that cutoff limits the upper alternation to the power supply voltage, 16V, and saturation, the lower alternation, is near 2V.

Exit Probe.

Use StmEd to Edit Vin to 2Vpeak

Reset the input stimulus to 2Vpeak, 1Vpeak base voltage, in preparation for the next exercise. Repeat the StmEd edit procedure used previously.

Exit StmEd when finished.

Connecting a Circuit Load

Using the Circuit Editor, connect the output coupling capacitor and load resistor by editing the netlist. Add these lines anywhere within the circuit description statements:

```
Co   4  7  10u
Rld  7  0  10K
```

Press Esc, and save the circuit file.

Calculating/Drawing the AC Load Line

In our studies of the common emitter amplifier we have been introduced to ac collector resistance (rc), the AC load line, and maximum swing of the AC load line. Maximum swing is defined as the smaller of ICQ(rc) or VCEQ-point. To make this maximum swing comparison we need to calculate the ac resistance of the collector circuit.

rc = RC || Rld
= 10K || 10K = 5KOhms

Then, collector swing limits are:

ICQ(rc) = (.75mA)(5K) = 3.75V
VCEQ-point = 7.5V

Therefore, maximum swing without clipping is limited to the smaller of the two values...3.75Vp.

Return to Figure 2 and draw the ac load line. The lower end of the ac load line is set by adding ICQ(rc)= 3.75v to VCE at the Q-point.

VCEQ + ICQ(rc) = 7.5V + 3.75V = 11.25V,

Draw the ac load line from 11.25V on the X_axis through the DC Q-point. Extend the line to the top of the graph.

Note: For optimum operation, the Q-point should, of course, be reset to the center of the AC load line to make ICQ(rc) equal VCEQ-point. But, for now, use PSpice to analyze the present Q-point setting.

Estimated AC Load Line Gain

With the load resistor connected, gain is reduced by one-half to an estimated rc / RE = 3.75. Since the maximum swing of the collector, without clipping, is limited to 3.75V, the maximum input signal at the base is approximately 1Vpeak (3.75Vp/1Vp= 3.75). Vin has been set at 2Vpeak, delivering 1Vpeak to the base of the transistor.

The next test will use transient_analysis to examine AC load line swing.

Run PSpice

Measuring Output Voltage, V(7)

At Probe title screen, select Transient_analysis, and Enter.
At Probe initial menu screen, select Add_trace, press F4, and Enter.
Arrow to V(7), the load voltage, and Enter twice.

The output swing is approximately 3.5Vpeak. There is some waveshaping of the output waveform that is caused by setting the input stimulus at maximum. This waveform distortion is evident on the positive alternation as the transistor is driven toward cutoff. Any further increase in the input would result in clipping. This clipping would, of course, occur on the positive swing of the output.

Note the offset voltage at 0V after the coupling capacitor.

Select V(4) to examine the collector swing, and to check output coupling efficiency.

Select Add_trace, Press F4, and Enter.
Arrow to V(4), and Enter twice.
V(4) and V(7) amplitudes are approximately the same.

Y_axis Range Adjustment

For more accurate measurements of V(4) and V(7) waveforms, display individually, and change the Y_axis range. For example, display V(4) in the following steps.

Select Remove_trace, and remove V(7), the load voltage trace.
Select Exit, and Enter, to return to the menu screen.

Collector voltage, V(4), is displayed.

Select Y_axis, and Enter.
Select Set_range, and Enter.
At Enter a range: type 5V 13V, and Enter.

The waveform amplitude is easier to read on this expanded "O'scope" scale.
Return the Y_axis to the original scale by using Auto_range:
Select Auto_range, and Enter.

The original Y_axis range is displayed.
Select Exit, and Enter, to return to Probe menu screen.

Measuring Base Input, V(2)

Input at the base of the transistor can be examined by first removing all traces, then selecting V(2).
Select Remove_trace, and clear the screen.
Select Add_trace, and display V(2).

The input signal swings from .7V to 2.7V, or 1Vpeak, as expected.

Calcualte AC Gain

Gain = v(out) / v(in) = V(7) / V(2) = 3.5Vp / 1Vp = 3.5
dB gain = 20 log (3.5) = 10.88

Examine Input vs Output Waveform Phase

V(2) is displayed from the previous section.
Select Add_trace, and display V(7).
Common-emitter output phase inversion is evident as the input and output waveforms are displayed at 180 degrees out-of-phase.
Select Exit, and Enter to leave Probe screen.
Select Exit_program, and Enter, to leave Probe program.

Overdrive the Input

To verify maximum input allowed, which is approximately 1Vpeak base voltage, use StmEd to increase the input stimulus to 2.5Vp (1.25Vp on the base), and examine V(7) for waveform clipping.

Repeat the StmEd procedure to edit the input stimulus to 2.5Vpeak.
Exit StmEd.

Run PSpice (To demonstrate overdriving)

At Probe title screen, select Transient_analysis, and Enter.
At Probe initial screen, select Add_trace, and display load voltage, V(7), and collector voltage, V(4).

Clipping of the output waveform demonstrates overdriving the amplifier. The AC load line illustrates maximum ac collector swing. The AC load line Q-point location nearer cutoff is indicated by the clipping of the positive alternation. When compared to the DC load line, the AC load line is rotated clockwise because of the reduced value of ac collector resistance. (See Fig. 2) Movement to cutoff is now only 3.75Vpeak (ICQ(rc)). Waveform clipping is occurring at cutoff before saturation. In fact, it would take a base input of 1.8Vpeak to drive the collector to saturation, a 7.5Volt swing.
(VCEQ-point= 7.5V)

StmEd Edit and Check for Saturation Clipping

Edit the base input to 1.8Vpeak to drive the transistor into saturation. Remember, for a base input of 1.8Vpeak, the input stimulus, Vin, must be set to 3.6Vpeak. Follow the procedures listed for StmEd editing.
Run PSpice and Probe Transient_analysis and examine the output waveform for distortion at saturation. Observe V(7).
Saturation clipping is evident.
Reset transient analysis input to 2Vpeak.

Decibel Gain Measurement

Set the value of Rs in preparation for dB gain measurements.
Edit the circuit file with the Circuit Editor:

Rs 1 9 2Ohms

Press Exc to save the edited circuit file.

Run PSpice

Select AC_sweep at Probe title screen, and Enter.
Select Add_trace, and Enter.

Type Vdb(7), and Enter.

Reset Y_axis range:
Select Y_axis, and Enter.
At Set_range: type 0 14 , and Enter.

Gain is measured at approximately 11dB, as calculated.
(dB gain = 20 log (3.5) = 10.88)

Exit Probe

Conclusion of Chapter 12

EOC Problems

(1) The Q-point should be set to the middle of the AC load line to achieve maximum ac collector swing before output waveform clipping. ICQ(rc) equals VCEQ-point at this setting. Decrease the value of RE to increase collector current to midpoint on the AC load line. Recalculate and draw the new AC load line. Edit the circuit netlist and Run PSpice to examine the new Q-point setting.

(2) Select a common-emitter amplifier in your theory text and write a circuit netlist. Complete all calculations for circuit behavior and run PSpice analysis to verify expectations. Use the Q2N2222A NPN model.

(3) Design a common-emitter amplifier. Write the netlist using the Q2N2222A NPN model and use PSpice to verify your design. VCC= 20V.

Notes:

Notes:

Chapter 13

COUPLING AND BYPASS CIRCUITS

.AC Sweep - Small-Signal Analysis

Objectives: (1) To evaluate the design of coupling capacitance
(2) To evaluate the design of bypass capacitance
(3) To use AC_sweep to evaluate frequency dependent capacitive circuits

Capacitors are used to isolate stages in systems and to short unwanted signals to ground. They are also used to provide "AC ground" to specific parts of circuits, thereby improving designs. In this exercise we will take a close look at capacitors and their associated circuits that are used to couple specific frequencies between stages and to bypass certain frequencies to ground. First, the coupling capacitor.

I. Coupling Capacitor Circuits

The capacitor is usually the first frequency dependent component studied in electronics. It is this dependence on frequency that makes it useful to the circuit designer. A capacitor is essentially an open circuit to DC; a short circuit to AC at higher frequencies. When using the capacitor to couple AC, the designer first identifies the lowest frequency to be moved from one part of a circuit to another, then calculates the capacitor size for transferring this frequency.

Any circuit can be said to be either a source or a load. (Fig. 1) Coupling capacitors are designed to move alternating current from sources to loads with a minimum of attenuation or waveshaping. Each source has output resistance (impedance) and each load has input resistance (impedance). Our coupling capacitor design must take these two impedances into consideration for optimum circuit performance.

We will use PSpice to simulate and analyze the circuit.

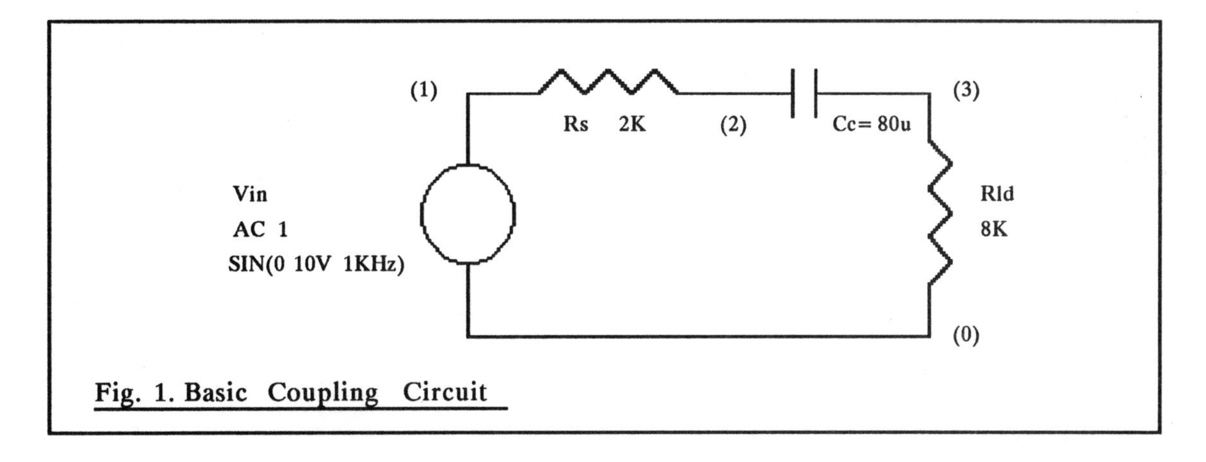

Fig. 1. Basic Coupling Circuit

The AC source in this test could, of course, be set to any frequency, but the adjoining circuit helps define just what frequency range will be coupled to the load, Rld. As frequency increases, capacitive reactance decreases, and more signal reaches the load. The circuit is essentially a high-pass filter with a minimum frequency.

In coupling capacitor circuits, the two frequencies of interest are defined as critical frequency and high-frequency border. Critical frequency is the frequency at which circuit rms current is 70.7 percent of maximum. Maximum current is available when capacitive reactance is zero. High-frequency border is 10X critical frequency. The capacitor to use for critical frequency coupling is determined by the equation,

Cc = 1 / (6.28)(Fc)(R) , defined;

Fc = critical frequency
R = the reactance of the capacitor equals the sum of the series resistance, Rs + Rld.

High-frequency border, Fh = 10(Fc).

The frequency to be coupled without significant attenuation is the high-frequency border. Rearranging the above equation to solve for capacitor size at Fh, we have,

Chf = 1 / (6.28)(Fh)(.1R)

With a high-frequency border of 2Hz, Rs= 2KOhms, and Rld= 8KOhms, the capacitance to insert into the circuit is,

Cc = 1 / (6.28)(2Hz)(1E3)
= 80uF

Enter the netlist and test the circuit with the 80uF capacitor.

Filename: COUPLCAP

```
*COUPLCAP - FREQUENCY RESPONSE
Vin 1 0 AC 1 SIN(0 10V 1KHz)
Rs 1 2 2K
Cc 2 3 80u
Rld 3 0 8K
*Control statements
.TRAN .02m 2m 0 .01m   ;transient analysis, 2Hz at 1KHz
*                      ;print_step=.02ms, final_time=2ms,
*                      ;delay=0, step_ceiling=.01ms
.AC DEC 5 100m 5k      ;AC_Sweep-small-signal analysis
*                      ;decade sweep, 5pts/DEC,
*                      ;start=.1Hz, stop=5KHz
.Probe
.End
```

Press Esc to exit the Circuit Editor and save the file.

Note: In PSpice, frequencies written with notation "m", like 100m in
this circuit file, denotes 100 millihertz, or .1Hz. If megahertz
is intended, Meg is written.

.AC Sweep - Small-signal analysis

The .AC sweep statement provides a small-signal frequency response analysis of the circuit. This control statement will sweep the circuit input, Vin in this exercise, with the frequencies specified. The input must include the AC sweep data: Vin 1 0 AC 1. The voltage input, "1" in this listing, can be set at any level but unity input will provide relative data that is more easily read. Probe sweep analysis program will display this data.

The input level, AC 1 for 1Volt, is insignificant since the circuit has been linearized for AC Sweep measurements. (Linearized means the AC Sweep analysis is performed on the linear component equivalent of the circuit) Circuit output amplitude, or any other node amplitude, is relative to (in proportion to) the input node, Node 1 in this particular circuit. Measurement results are valid where circuit frequency limits are not exceeded and circuits are not driven into nonlinear operation.

PSpice performs one of three types of frequency definitions in the .AC sweep statement;

.AC {sweep type: DEC LIN OCT} points start_frequency end_frequency

They are best seen with examples:

.AC DEC 5 100m 5k. A logarithmic sweep, by decades, is requested with 5 data point measurements per decade as the first number in the statement. Decade is a 10X increase in frequency. Decade steps: .1Hz, 10Hz, 100Hz, 1KHz, 10KHz. The range: .1Hz - 5KHz.

.AC LIN 200 10K 1MEG, defines a linear sweep with 200 data points during the sweep. Range: 10KHz - 1MEGHz.

.AC OCT 10 10K 1MEG, defines a logarithmic sweep, by octaves, with 10 data points per octave. Examples of octave increases in this statement are 10KHz, 20KHz, 40KHz, 80KHz,...etc. The sweep range is 10KHz -1MEGHz.

Viewing the Input Stimulus for Transient_Analysis

We will take a look at the 1KHz input stimulus with the Stimulus Editor before running circuit frequency analysis. Select StmEd, Edit, and Enter.

The first screen displays the two cycles at 1KHz that was written in the Vin statement and the analysis time, 2ms, in the transient analysis control statement. Vin amplitude is 10Vpeak. StmEd allows the user to change the input waveform by selecting Modify_stimulus, followed by Transient_parameters. All changes are automatically written to the circuit file netlist and can be viewed with the Circuit Editor.

Select Exit, and Enter, to leave StmEd's initial screen.
Select Exit_program, and Enter, to leave StmEd.

Run PSpice

Select Analysis, Enter, Run PSpice, and Enter.
The Probe title screen will be displayed when .AC Sweep and .TRAN analyses are completed.

Probe "Oscilloscope"

The two Probe options for examining the coupling capacitor network are AC_sweep and Transient_analysis. AC_sweep provides the input sweep Vin, at Node 1, over the frequency range set in the .AC control statement: .1Hz to 5KHz. Again, 1Volt input for this sweep is set by Vin 1 0 AC 1. Analysis output is written to the data file, COUPLCAP.DAT. Transient_analysis in Probe will also use the data file to display the results of the analysis of 2 hertz at 1KHz. This stimulus is set by the following underlined section of the input statement;

Vin 1 0 AC 1 SIN(0 10V 1KHz)

Vin defines (1) the type of waveform...sine wave, (2) the offset voltage...0V, (3) the input level...Vpeak= 10V, and (4) the frequency to be analyzed...1KHz.
The transient analysis control statement is listed as,

.TRAN .02m 2m 0 .01m

The .TRAN statement defines the transient time period in which to analyze the operation of the circuit; final_time= 2ms (2 hertz at 1KHz). It also sets print_step with the .02ms entry, which is used for .print and .plot outputs. Typically, print_step is set to 1/100th of final_time. Results_delay is set to 0, for no delay. Step_ceiling is limited by the last entry in this statement, .01ms. The default value would have been final_time/50 = .04ms. A .01ms step_ceiling will provide better Probe graph presentations.

AC_sweep

What are we expecting AC_sweep to confirm about our circuit? Our original design calls for coupling all frequencies above 2Hz to Rld. If we have calculated correctly, the coupling capacitor should approximate a short at all frequencies above 2Hz. All frequencies below 2Hz should be attenuated.

Select AC_sweep, and Enter.

The initial graph has frequency displayed on the X_axis: 100mHz to 10KHz. Our circuit program sweep frequency spans 100mHz to 5KHz. The graph covers the area of interest...2Hz. Now, let's look at V(3), the swept output of the coupling circuit.

Select Add_trace, and Enter.

Press F4 for a list of variables and expressions available to view in AC_sweep.

Select V(3), and Enter twice.

The AC_sweep screen displays a steady output of 800mV at all frequencies above our design break of 2Hz. All frequencies above 2Hz are passed by the coupling capacitor to the load without significant attenuation.

Note: At frequencies where Cc is a short, Rs and Rld act as a voltage divider, V(3) = (Rld / (Rld + Rs))(Vin=1V). V(3) maximum = 800mV.

While at this screen, take a look at the voltage at V(2), the node on the source side of the capacitor.

Select Add_trace, Enter, F4, arrow to V(2), and Enter twice.

At V(2), we can see the capacitor attenuated the input from DC to near 2Hz, then, as reactance decreased, its response gradually became a short and the voltages at V(2) and V(3) are the same...800mV.

Viewing I(Cc)

The next few steps will familiarize the user with removing selected traces. Our next objective is to take a look at current through the coupling capacitor, I(Cc). V(3) and V(2) are presently displayed.

Select Add_trace, Enter, F4, arrow to I(Cc), and Enter twice.

As presently displayed with V(3) and V(2), I(Cc) cannot be determined. Let's remove the two voltage traces.

Select Remove_trace, and Enter.

Arrow to the Select option to choose the traces to be removed, and Enter.

As the screen prompts state, use the Arrow keys to move the highlight to the trace to be removed, then press the Spacebar to select that trace. Repeat these steps for both V(3) and V(2). Press Enter. I(Cc) should be the only trace on the screen, and the Y_axis displays current, 40uA to 100uA.

Exit the Remove_trace screen.

Notice the shape of the capacitor current trace is identical to V(3), the output voltage trace that we observed earlier. Current maximizes at 100uA at approximately 2Hz and remains constant at frequencies above 2Hz.
(Imax= Vin / (Rs + Rld)= 1V / (2K + 8K)= 100uA)

For practice, use Add_trace to display all of the current traces available on the same screen with I(Cc). Since this is a series circuit, all traces indicate the same current.

Select Remove_trace and by individually selecting a trace, remove one at a time for practice in removing traces.

Exit the Remove_trace screen.
Select Exit, and Enter, to leave the AC_sweep part of the Probe program.

Probe's title screen should be displayed.

Transient_analysis

Let's look at the 1KHz sine wave after it has passed through our coupling circuit. We do not expect attenuation of this signal since it is above the circuit's high frequency border of 2Hz.

Select Transient_analysis, and Enter.
Notice the X_axis in transient analysis is Time, 0s to 2.0ms, covering the time specified in the control statement.
Select Add_trace, Enter, F4, arrow to V(3), and Enter twice.

V(3) is an undistorted sine wave, as expected, and since the coupling capacitor is a short to the 1KHz signal, the amplitude is maximum...8Vpeak. Remember, the input amplitude for transient analysis was set at 10Vpeak and the resistor divider network provides 80% of input voltage to the output.
(Rld / (Rld + Rs))

Select Exit, and Enter, to leave the Transient_analysis section of the Probe program.
Select Exit_program, and Enter, to leave Probe.

II. Bypass Capacitor Circuits

Bypass circuits are used to couple unwanted frequencies to ground. In other designs they provide "AC ground" for better circuit performance. In both type circuits, the resistance associated with the bypass capacitor must be defined and used in calculations. But, unlike the coupling capacitor, the resistance is in parallel instead of in series with the bypass capacitor. Bypass designs can be considered as low-pass filters in the circuit, since higher frequencies are passed to ground.
Review the test circuit in Figure 2.

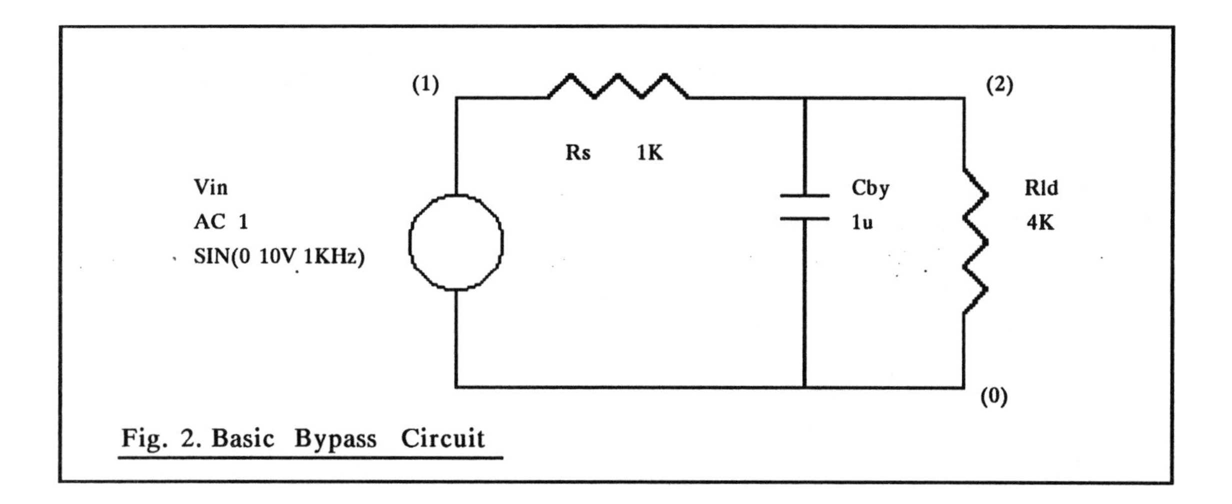

Fig. 2. Basic Bypass Circuit

In bypass capacitor circuits, the two frequencies of interest are again defined as critical frequency and high-frequency border. Critical frequency is the frequency at which the capacitive reactance is equal the parallel resistance, Rs || Rld, in the above circuit. High-frequency border is 10X critical frequency. At this frequency the capacitor is a short and provides an AC ground at Node 2. All frequencies above high-frequency border are also "passed" to ground.

The equation to use when solving for bypass capacitor size at a specific critical frequency is,

Cc = 1 / (6.28)(Fc)(R) , defined;

Fc = the critical frequency.
R = the reactance of the capacitor equals the parallel resistance sum R, R= Rs || Rld.

High-frequency border, Fh = 10(Fc).

Rearranging the equation to solve for bypass capacitor size at a specific high-frequency border, we have,

Chf = 1 / (6.28)(Fh)(.1R),

where the value of R is one-tenth the parallel sum of Rs || Rld.

With a high-frequency border of 2KHz, Rs= 1KOhms, and Rld= 4KOhms, the capacitance to insert into the circuit is,

Chf = 1 / (6.28)(2KHz)(80 Ohms)
= 1uF

Enter the following netlist at the Circuit Editor.

Filename: BYCAP

```
*BYCAP - FREQUENCY BYPASS CAPACITOR
Vin  1 0 AC 1 SIN(0 10V 1KHz)
Rs   1 2 1K
Rld  2 0 4K
Cby  2 0 1u
.AC DEC 5 100m 10k          ;decade sweep
*.AC LIN 250 100m 10k       ;insert later for linear sweep
.Tran .02m 2m 0 .01m
.Probe
.End
```

Press Esc to exit and save the Circuit File.

Run PSpice

At Probe

Select AC_sweep, and Enter.

Select Add_trace, Enter, F4, arrow to V(2), and Enter twice.

Frequency is displayed on the X_axis. The bypass capacitor begins to pass all frequencies above 20Hz to ground. At high_frequency border, 2KHz, and above, the bypass capacitor is essentially a short.

Note the 1KHz point on the circuit's frequency response graph.

Select Exit, and Enter, to leave AC_sweep.

Select Transient_analysis, and Enter.

Select Add_trace, Enter, F4, V(2), and Enter twice.

The 1KHz input signal is not above the circuit's high-frequency border design and is somewhat attenuated.

Select Add_trace, and display V(1) with V(2).

Note attenuation and the circuit's initial response and waveform phase shift.

Select Exit, and Enter.

Select Exit_program, and Enter, to return to Control Shell.

Remove the decade sweep line from the netlist and insert the linear sweep line. Run PSpice and repeat .AC sweep for the same measurements.

Display the sweep input, V(1), and V(2) simultaneously. Study the graph.

Exit Probe to Control Shell.

Conclusion of Chapter 13

EOC Problems

(1) Design several coupling circuits with various critical frequencies.
Write the netlists and Run PSpice to verify your calculations.

(2) Design several bypass circuits with various critical frequencies.
Write the netlists and Run PSpice to verify your calculations.

Chapter 14

RESONANCE; SERIES AND PARALLEL

RC - LR PHASE SHIFT CIRCUITS

COMPLEX NUMBERS IN AC CIRCUITS

Objectives: (1) To examine series resonance using PSpice
(2) To examine parallel resonance using PSpice
(3) To measure RC and LR phase shift circuits
(4) To introduce PSpice phase and magnitude measurements

The study of electronic fundamentals can be rather dull until the chapter on resonance. In earlier chapters, components seem to be dormant independent responders to whatever prods them. At resonance, the circuit explodes into interaction. Energy sloshes back and forth and through the entire design, and almost every node is frequency sensitive. Each reactive component seems to recognize every other reactive component, even those hidden inside special devices. And if frequency is high enough, circuits seem to create capacitors and inductors all over the place. From this chapter forward, there is no reason for being bored. The student, himself or herself, can resonate with and receive from the electronic phenomenon. The excitement is on. In this exercise, we will examine small circuits in resonance and use PSpice and Probe to observe the activity.

I. Series Resonance

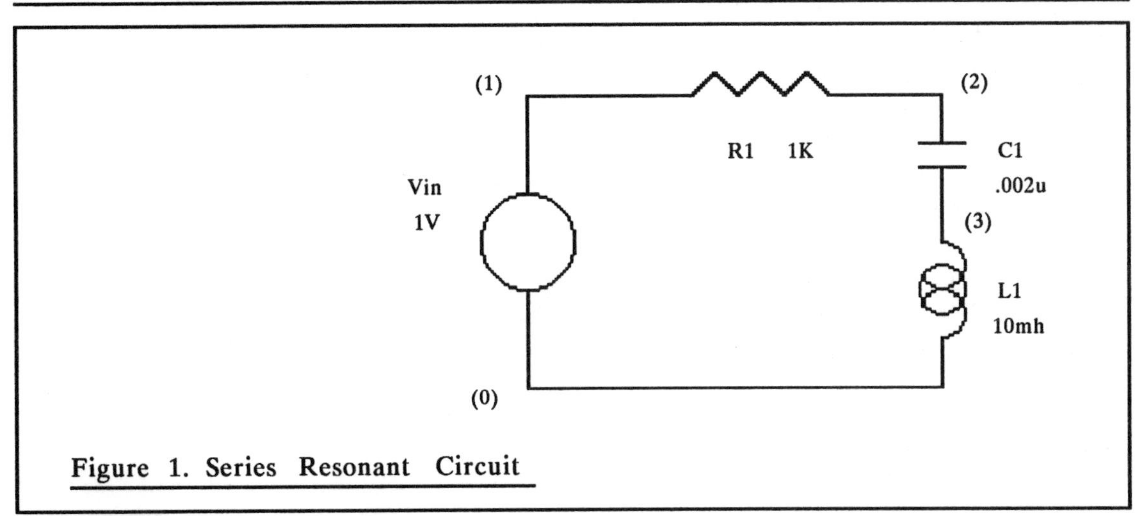

Figure 1. Series Resonant Circuit

The Netlist

Enter the following circuit description and control statements with the Circuit Editor. Select Files, Enter, Current File, and Enter.
Filename: SRESONANT
Select Edit, and Enter.

```
*SERIES RESONANT LC CIRCUIT
Vin  1  0  AC  1      ;AC input  voltage
R1   1  2  1K
C1   2  3  .002u
L1   3  0  10mh
*Control statements
.AC  LIN  200  5K  150K
.End
```

Press Esc and save the circuit file.

Netlist Description and Control Statements

A netlist defines the circuit and the analyses to be performed. Description and control statements are used to describe this information in PSpice. PSpice converts the netlist into a computer program.

Description statements usually appear first in the circuit netlist but this is not a must. Control statements usually follow. PSpice is not particular where in the netlist a particular line is located except for first and last lines. These two lines are reserved for the title line and the .End statement.

Control statements are used to tell PSpice what you want analyzed in the circuit. For example:

```
.AC  LIN  200  5K  150K
```

This line provides a linear sweep of the input from 5KHz to 150KHz, with 200 data points in the sweep. A small-signal circuit response at specific frequencies is the measured results of this analysis. For example, circuit resonance is measured using this control statement.

.End

The .End control statement indicates the last line in the circuit netlist. PSpice begins simulation and analysis only when there is an .End statement. Another circuit netlist can be entered after the .End statement and must close with .End.

Circuit Expectations

Capacitive reactance and inductive reactance has been explored in theory studies. The 90 degree phase shift of current and voltage in inductors and capacitors should be understood before moving ahead in electronics. These components are frequency sensitive and phase shift responsive. And, the question might be asked, what would happen to circuit energy if reactance related components are soldered together and examined with frequency? One answer to that question is interesting.

As frequency increases, inductors react more and capacitors react less, and at one specific frequency, in series LC circuits, the reactances are equal and cancel. At this frequency, the circuit "resonates" and, in series circuits, resistance controls current. At this frequency, circuit voltages and currents respond in what is called "resonant rise" and "resonant fall." It is not only interesting but the possibilities are phenomenal. Radio and TV receiver frequency selection is just one use for these resonant circuits.

With PSpice, the test circuit is swept with a range of frequencies much like a sweep generator. Circuit simulation examines all node voltages and component currents, including the voltage source. With Probe, circuit frequency response curves, including quality of frequency selection (Q), can be drawn for each node of interest.

Calculate Xl, Xc, resonant frequency and current for the circuit in Figure 1. Compare your estimates with the values in Table I.

Estimated Values		PSpice Values
Input Voltage	= 1V	
Resonant frequency	= 35.6 KHz	Measured = ____ KHz
At Resonance:		
Inductive reactance	= 2236 Ohms	
Capacitive reactance	= 2236 Ohms	
Resistance, R1	= 1 KOhm	
Current	= 1 mA	Measured = ____ mA
Voltage across L, V(3)	= 2.236 V	Measured = ____ V
Voltage across C, V(2,3)	= 2.236 V	Measured = ____ V
Voltage across LC, V(2)	= ~0 V	Measured = ____ V

Table I. Series Resonant Circuit Values

R1 will be edited to .1 ohms later in the exercise and V(3) will be measured again for resonant rise in voltage across L1.

V(3), voltage across L1, =______V (Resonant rise with R1= .1 Ohms)

Run PSpice

Circuit Current

Select Add_trace, and Enter. Press F4 for a list of values to display.
Arrow to either current and Enter twice.
Record resonant current value in Table I.
Sketch and label graphs of the following Probe displays.

The X_axis frequency scale includes the range specified in the AC sweep statement. Current is displayed on the Y_axis. Note that current is maximum at resonance and is controlled by the 1KOhm resistor (I= 1V/1K= 1mA).

Reactance, before and after resonance, is the phasor sum of Xl and Xc. The out-of-phase currents and voltages in the inductor and capacitor have produced a zero reactance at one frequency,...resonant frequency. The width of the response curve plotted by Probe is determined by the Q of the circuit, which is also controlled by the resistor. Estimate the bandpass.

Again, impedance at resonance is resistive, since reactances cancel. Purely resistive impedance aligns circuit current in-phase with the input stimulus, Vin. At resonance, voltage across the resistor is maximum, and minimum across the LC circuit at V(2). Viewing V(2) will give some indication of zero reactance at resonance.

Select Remove_trace, and clear the screen.

Circuit Voltages

Select Add_trace, and display V(2), the voltage across the LC combination. Record V(2) voltage at the resonant frequency in Table I.

The measured voltage across the LC network drops from near maximum, where the sweep is off resonance, to near 0Volts at resonance. At resonance, most of the circuit input voltage is across R1.

Reset X_axis to measure resonant frequency.

Select X_axis, and Enter. Select Linear, and Enter.

Select Set_range, Enter, and type 34Kh 36Kh , and Enter.

Record the measured resonant frequency in Table I.

Select Auto_range, Log sweep, Exit, Remove_trace, and clear the screen.

Select Add_trace, and display V(3), the resonant rise in voltage across the inductor. Record this voltage in Table I.

Select Remove_trace and clear the screen.

Select Add_trace, and measure the resonant rise in voltage across the capacitor.

Type V(2) - V(3), and Enter.

Record this value in Table I.

Select Remove_trace, and clear the screen.

Minimum Circuit Resistance, High Q Measurement

The following exercise is presented to illustrate series resonant circuit quality, Q, or figure of merit. Q is a measure of the ratio of reactance to series resistance at resonance. Higher Q means higher resonant rise in voltage and greater sharpness of frequency tuning.

The Probe Graphical Waveform Analyzer has a viewing option that is sometimes difficult to duplicate at the student Lab workbench...a minimum circuit resistance that produces a high circuit Q and an extremely high resonant rise in inductor and capacitor voltages. As an example, return to Control Shell and, using the Circuit - Devices option, edit R1 to .1Ohms.

Exit Probe.

At Control Shell

Select Circuit pop-down menu, and Enter.

Select Devices, and Enter. Arrow to R1, as "Device to Change," and Enter. Arrow to "resistance = 1.000k," and Enter.

For "new value?", type .1Ohms, and Enter.

Press Esc as needed to exit.

PSpice has edited this new value for R1 into the circuit netlist. Return to the Circuit Editor to examine this editing.

Press Esc to exit.

Run PSpice

High Circuit Q Voltage Measurement

At Probe

Select Add_trace, and complete the procedure to display V(3), the resonant rise in voltage across the inductor.

The sharper response curve at approximately 36KHz is the result of lower resistance in the circuit. Note the 1.2KVolts measured at V(3). Record this voltage in the space provided just below Table I, on page 3.

Select Remove_trace, and clear the screen.

Select Add_trace, and display V(2), LC network voltage.

As expected, the out-of-phase voltages across L1 and C1 cancel, and V(2) is minimum at resonance. Next, measure the voltage across C1.

Select Remove_trace, and clear the screen.

Select Add_trace, and Enter.

Type V(2) - V(3), and Enter.

The graph displays the same 1.5KV level that was measured across L1. Again, these two voltages are out-of-phase and cancel.

Select Remove_trace, and clear the screen.

High Q Circuit Currents

Select Add_trace, and display all current measurements simultaneously; I(L1), I(C1), I(R1), and I(Vin). Series current is the same throughout the circuit. At resonance, current is measured at approximately 530mA.

Exit Probe to Control Shell.

This concludes the series resonant circuit exercise.

We will experiment with parallel resonance in the following section.

II. Parallel Resonance

Previous studies have theorized that parallel resonant circuits have maximum impedance and minimum current. This theory will be tested in this section after

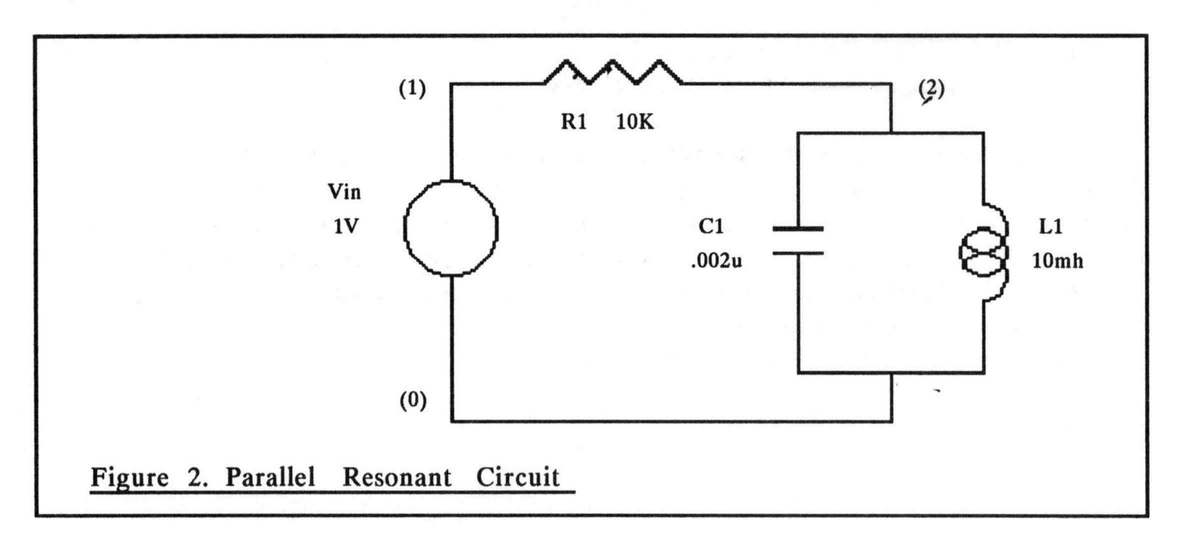

Figure 2. Parallel Resonant Circuit

making circuit estimates. Evaluate the circuit in Figure 2 by first calculating Xl, Xc, and resonant frequency. Compare you values with those in Table II.

Enter the netlist. Filename: PRESONANT

```
*PARALLEL RESONANT LC CIRCUIT
Vin  1  0  AC  1
R1  1  2  10K
L1  2  0  10mh
C1  2  0  .002u
*Control statements
.AC  LIN  200  5K  150K
.End
```

Estimated Values		PSpice Values
Input Voltage	= 1 V	
Resonant Frequency	= 35.6 KHz	Measured = ____KHz
At resonance:		
Inductive reactance	= 2236 Ohms	
Capacitive reactance	= 2236 Ohms	
Resistance, R1	= 10 KOhm	
Circuit current	= Minimum	Measured = .375 uA
C and L current	= Maximum	Measured = 450 uA
Voltage across LC, V(2)	= Maximum	Measured = 1 V

Table II. Parallel Resonant Circuit Values

Run PSpice

At Probe

Circuit Voltage

Select Add_trace, Enter, and press F4.
Arrow to V(2), parallel LC voltage, and Enter twice.

The response curve measuring voltage across the LC (tank) circuit is displayed. Note the swept frequencies on the X_axis. At resonance, all of the input voltage is measured at V(2). The width of the curve is circuit frequency response. This width is determined by the Q of the circuit. Estimate circuit Q.
Sketch and label graphs of Probe displays in these test measurements.
Record V(2) voltage at resonance in Table II.
Select Remove_trace, and clear the screen.

Circuit Current

Current should be minimum at resonance, with the same response curve shape as V(2), but inverted.
Select Add_trace, and display I(R1), circuit current.
The current response curve for the parallel resonant circuit is displayed. Current drops to 0A from a maximum of 100uA. Record resonant current amplitude in Table II. Sketch and label the graph. Next, reset X_axis range to measure resonant frequency.
Select X_axis, Enter, select Linear, and Enter.
Select Set_range, and Enter.
At Enter a range: type 34Kh 36Kh , and Enter.
Record resonant frequency in Table II. Select Auto_range, and Enter.
Select Log sweep, and Enter.
Select Exit, and Enter.
Select Remove_trace, and clear the screen.

Next, measure LC parallel (tank) current for resonant rise.
Select Add_trace, and display I(L1) and I(C1).
Maximum current is approximately 450uA. Record this value.
Select Remove_trace, and clear the screen.

Display all currents, I(L1), I(C1), I(R1), and I(Vin) on the same screen for a colorful, informative view of a parallel LC circuit in series with resistance.
Exit to the Control Shell.

Conclusion of parallel resonant circuit exercise.

In the next section, we will examine phase shift circuits and conclude the chapter with an introduction to PSpice measurements of complex numbers.

Resonant Circuit Problems

(1) Design a LC series resonant circuit and enter the netlist. Sweep the input to cover test frequencies. Run PSpice to verify your estimations.

(2) Design a LC parallel resonant circuit and enter the netlist. Sweep the input to cover test frequencies. Run PSpice to verify your estimations.

(3) Select two or three LC resonant circuits from your theory textbooks for analysis. Complete all calculations, enter the netlist, and Run PSpice to evaluate the designs.

Notes:

III. RC And LR Phase Shift Circuits

In theory studies we have learned that current leads voltage by 90 degrees in a capacitor. Also, in an inductor, voltage leads current by 90 degrees. PSpice AC_sweep analysis measures phase shift in capacitive and inductive circuits and presents phase data in the Output file. An RC series phase shift circuit will be analyzed in this section, followed by an LR parallel phase shift circuit.

Transient_analysis has been included to allow Probe to present an "O'Scope" current/voltage phase shift display of circuit phase differential waveforms.

This will be the first opportunity to write two circuit netlists in one circuit file and to make the necessary choices when using transient_analysis presentations in Probe. It is also the first time to request phase measurements, VP(C1), VP(R1), and IP(R1), along with node differential voltages, i.e., V(1,2). Write the following circuit file including both the RC circuit and the LR circuit and compare each with respective circuit diagrams in Figures 3 and 4. Include comments.

Filename: RCLR

```
*RC PHASE-SHIFTER CIRCUIT
Vin 1 0 AC 1 SIN(0 100V 60Hz 0 0 0)
C1 1 2 .05u
R1 2 0 53K
.Tran .66ms 33.3ms 0 .33ms
.AC LIN 1 60Hz 60Hz
.Print AC VP(1) V(1,2) VP(C1) V(2) VP(R1) I(R1) IP(R1)
.Probe                  ;VP(C1)= C1 voltage phase angle in degrees
*                       ;VP(R1)= R1 voltage phase angle in degrees
.End                    ;IP(R1)= R1 current phase angle in degrees
*LR PARALLEL PHASE-SHIFTER CIRCUIT
Vs 4 0 AC 1 SIN(0 1V 1KHz 0 0 0)
R2 4 0 500Ohms
R3 4 5 5Ohms
L1 5 0 80m
.Tran .02ms 2ms 0 .01ms
.AC LIN 1 1K 1K
.Print AC V(4) VP(4) I(R2) IP(R2) I(L1) IP(L1)
.Probe
.End
```

Press Esc to save the circuit file.

RC circuit phase angle is determined by capacitive reactance and resistance. Both are 53KOhms in this circuit, making the phase angle= -45 degrees. Calculated:

```
TAN(ANGLE)= -Xc / R
          = -53K / 53K = -1
ANGLE     = -45 degrees
```

Any combination of R and C can be used to shift circuit phase by any specific number of degrees. Node 2 output is the phase shifted voltage. The phasor diagram for the RC circuit in this experiment is presented in Figure 5.

The Test Circuits

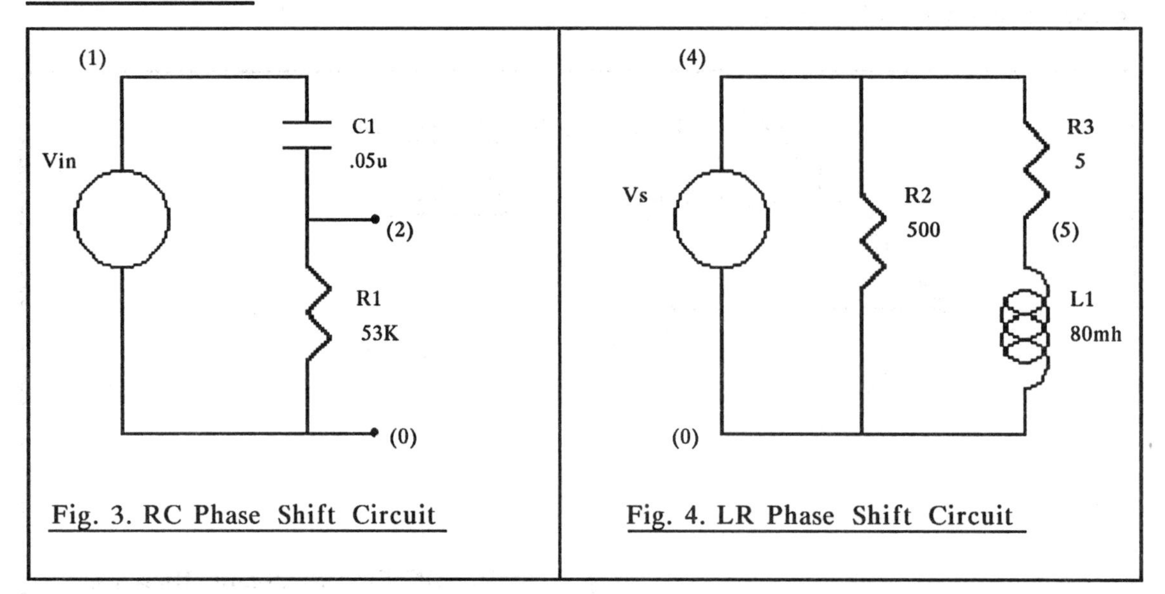

Fig. 3. RC Phase Shift Circuit

Fig. 4. LR Phase Shift Circuit

Note that capacitor voltage is always lagging resistor voltage by 90 degrees in a series circuit.

Next, PSpice analysis and Probe will verify calculations.

Run PSpice

At Probe

Select Transient_analysis, and Enter.
At "List of Analysis Sections" screen, arrow to Select_sections, and Enter.
Arrow to RC PHASE-SHIFTER CIRCUIT, and press Spacebar.
Press Enter.
At Probe menu screen, select Add_trace, and Enter.
Press F4, and arrow to V(1), the input voltage, and Enter twice.
Repeat for V(2), circuit output voltage.

The output waveform is delayed by an estimated 2.1ms. Measure the delay near the center of the graph, on the 0V line.

Calculate phase shift:

Phase shift= (2.1ms / 16.67ms)(360 deg.)= approximately 45.4 degs.

Exit Probe.

PSpice analysis has recorded the exact circuit response to the 60Hz input stimulus in the Output file.

Browse Output

The requested voltages, currents, and phase measurements are displayed in the Output file. Move down to the first AC Analysis screen and note that capacitor voltage, V(1,2)= .7075V, has been shifted, VP(C1)= -45 degrees. The output voltage, V(2)= .7068V, is measured at VP(R1)= +45 degrees. Page down for the last two measurements. I(R1) measures current at .0133mA. IP(R1) measures phase at +45 degrees.

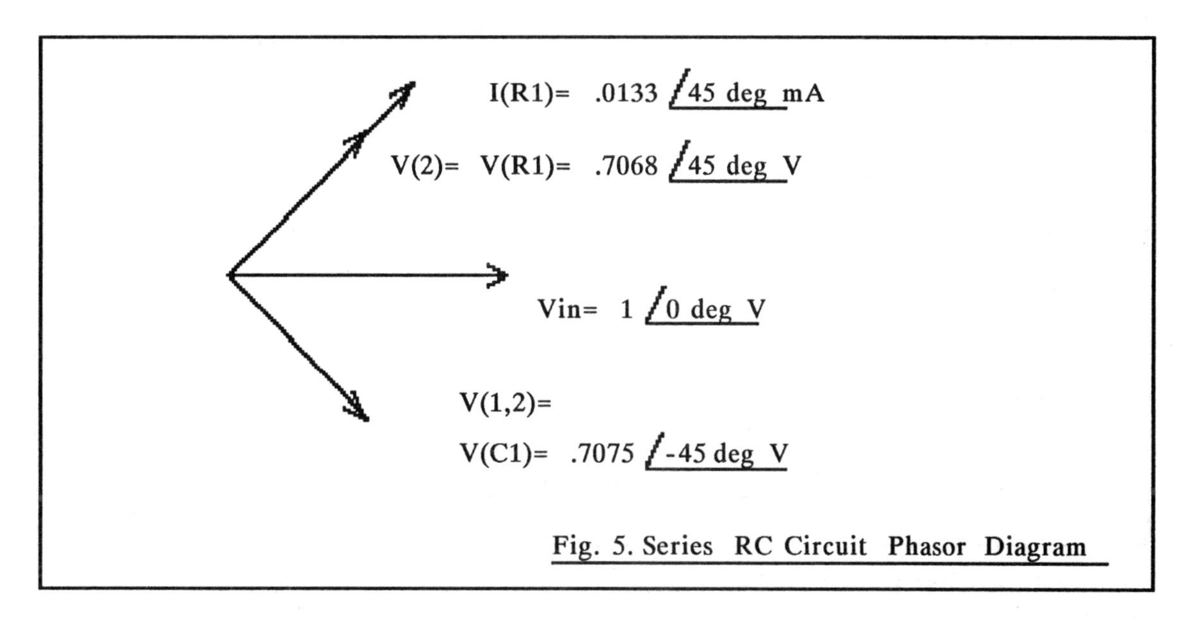

Fig. 5. Series RC Circuit Phasor Diagram

LR Phase Shift Circuit

The parallel LR circuit will be analyzed in this section. Inductive reactance is approximately equal to parallel resistance in the circuit in Figure 4. Branch currents should be equal, but L1 current phase angle should reflect the inductive branch.

Check the circuit netlist before running PSpice.

Run PSpice

At Probe

Select Transient_analysis, and Enter.

At "List of Analysis Sections" screen, arrow to Select_sections, and Enter.

Arrow to LR PARALLEL PHASE-SHIFTER CIRCUIT, and press Spacebar, then Enter.

At Probe menu screen, select Add_trace, and Enter.

Display V(4) and V(5).

The traces represent 2Hz of the 1Vp, 1KHz, stimulus.

Select Remove_trace, and clear the screen.

Select Add_trace, and display I(R2), the resistive branch current.

Repeat for I(L1), the inductor branch current.

Branch currents have approximately the same peak amplitude, 2mA, with the current through L1 lagging the current through R2 by approximately -90 degrees, as expected. The Output file will present more accurate measurements.

Exit Probe.

Browse Output

Select Browse Output and page down past the LR netlist to the first AC Analysis screen.

Input stimulus voltage and current are in-phase. The resistive branch current, I(R2), is 2mA and in-phase, IP(R2), with the input stimulus. Inductive branch current, I(L1), is approximately 2mA and near the -90 degree phase angle expected, IP(L1). See phasor diagram below.

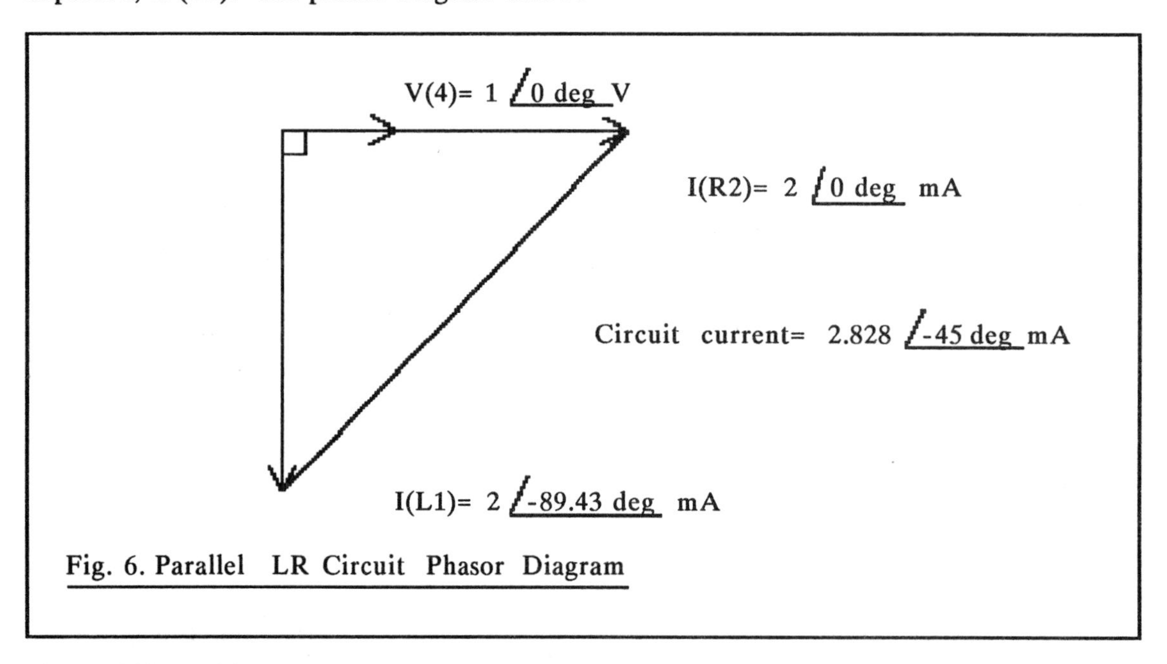

Fig. 6. Parallel LR Circuit Phasor Diagram

Phase Shift Problems

1. Sketch a parallel RC experiment using equal reactance and resistance. Complete all calculations and Run Spice to verify current, voltage, and phase estimates.

2. Sketch a series LR experiment using equal reactance and resistance. Complete all calcualtions and Run PSpice to verify current, voltage, and phase estimates.

3. Use PSpice to analyze RC, LR, and RCL circuits in your theory textbooks.

IV. Complex Numbers in Series Circuits

Real and imaginary circuit values can be entertaining using PSpice experimentation. The real component and imaginary component of current in the series circuit below in measured and presented in the Output file using AC_sweep analysis in PSpice. Impedance is calculated and phasors are drawn to illustrate circuit behavior.

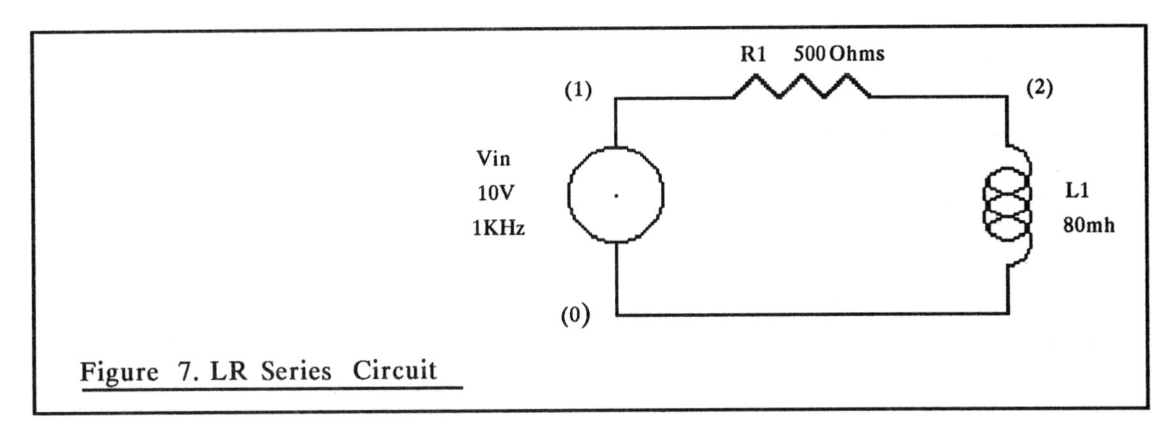

Figure 7. LR Series Circuit

The reactance of the inductor is 500 ohms, the same value as the resistance in the circuit. Net impedance is 707 ohms. Total current will be 14.14mA, part a real component and part an imaginary component, j. Current in the inductor will lag voltage by -90 degrees, but the resistive component in the circuit will limit circuit current phase shift to -45 degrees.

PSpice will be programmed to display R1 current, I(R1), the real component of current in R1, IR(R1), and the imaginary component of current in R1, II(R1). Current phase, studied in the previous section, will be measured and displayed in the Output file as IP(R1). Enter the LR netlist. We will return later and use the RC netlist.

RL Circuit

Filename: REALIMG1

```
*LR REAL/IMAGINARY  MEASUREMENTS
Vin  1  0  AC  10V
R1   1  2  .5K
L1   2  0  80m
.AC  LIN  1  1K  1K
.Print  AC  I(R1)  IR(R1)
.Print  AC  II(R1)  IP(R1)
.End
```

RC Circuit

Filename: REALIMG2

```
*RC REAL/IMAG  TEST
Vin  1  0  AC  10V
R1   1  2  1000
C1   2  0  .16u
.AC  LIN  1  1K  1K
.Print  AC  I(R1)  IR(R1)
.Print  AC  II(R1)  IP(R1)
.End
```

Press Esc and save the circuit file.

Run PSpice

Browse Output File

Page down to AC Analysis. I(R1) current is listed as .0141A. The real component, IR(R1), is near .01A, and the imaginary component, II(R1) is -.01A. Phasor addition of real and imaginary components is .0141A. Note that IP(R1) phase is at -45 degrees as expected.

Impedance, Z, calculated from R1 and XL in the circuit is 707 ohms. Impedance angle is the angle with a tangent XL/R1 = 1, or 45 degrees.

Using Z= Vin/I(R1), impedance is the same, 707 Ohms, at 45 degrees.

Z= Vin / I(R1)

= 10 ∠0deg V / .0141 ∠-45 deg A

= 707 ∠45deg Ohms

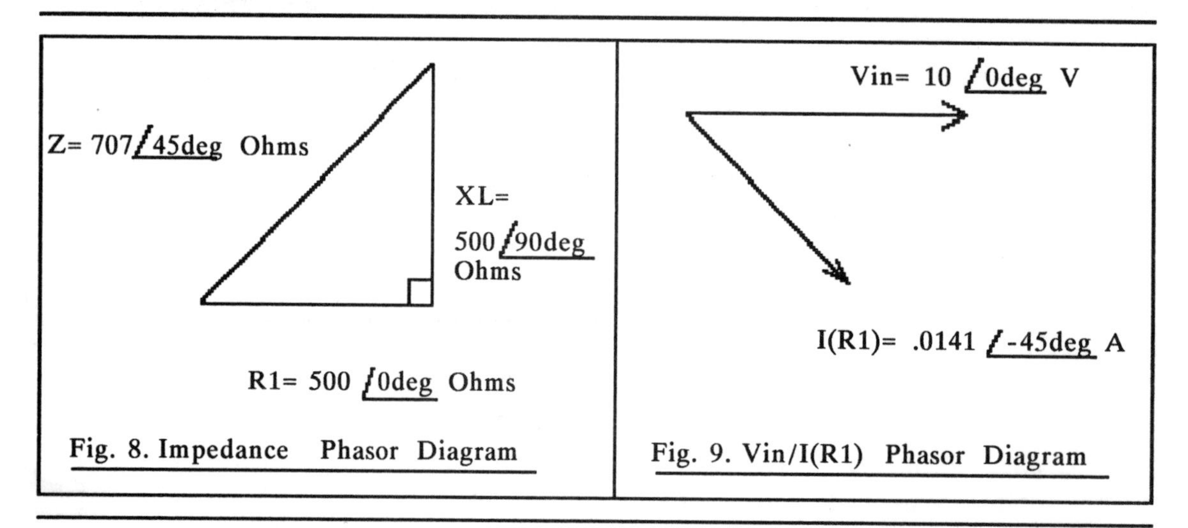

Fig. 8. Impedance Phasor Diagram

Fig. 9. Vin/I(R1) Phasor Diagram

RC Series Circuit

Sketch the circuit and repeat the above experiment and draw phasor diagrams for the RC netlist on page 13. Capacitive reactance is 1000 Ohms, matching the resistive value, for a phase shift of 45 degrees.

Impedance, Z, calculated from R1 and XC is 1414 Ohms.

Using Z= Vin/I(R1)= 10V/.007A= 1414 Ohms.

Output file data: I(R1)= .00709A IR(R1)= .00503A,
II(R1)= .005A IP(R1)= 44.85 degrees.

Complex Number Problems

1. Use PSpice to verify RC and L complex number solutions in theory textbook problems.

Chapter 15

SERIES REGULATOR

Objectives: (1) To use PSpice to simulation a series voltage regulator
(2) To use PSpice to evaluate series regulator design changes

In the study of rectifiers, a toleration of 10% ripple in output voltage is sometimes allowed. The idea being that, if necessary, other circuit components will provide additional filtering of the less than perfect DC voltage. In some more critical designs, regulators are inserted to remove ripple and other voltage variations. A series regulator is examined in this chapter to observe its operation and regulation capabilities.

The regulator in Figure 1 will be evaluated. Testing will begin by changing the size ratio of R3 and R4, the feedback divider circuit. The effects of these changes on closed-loop gain will be calculated, then PSpice analysis will be used to confirm estimates. In the last part of the exercise, the input will be swept with a known percentage change and circuit responses to these changes will be measured.

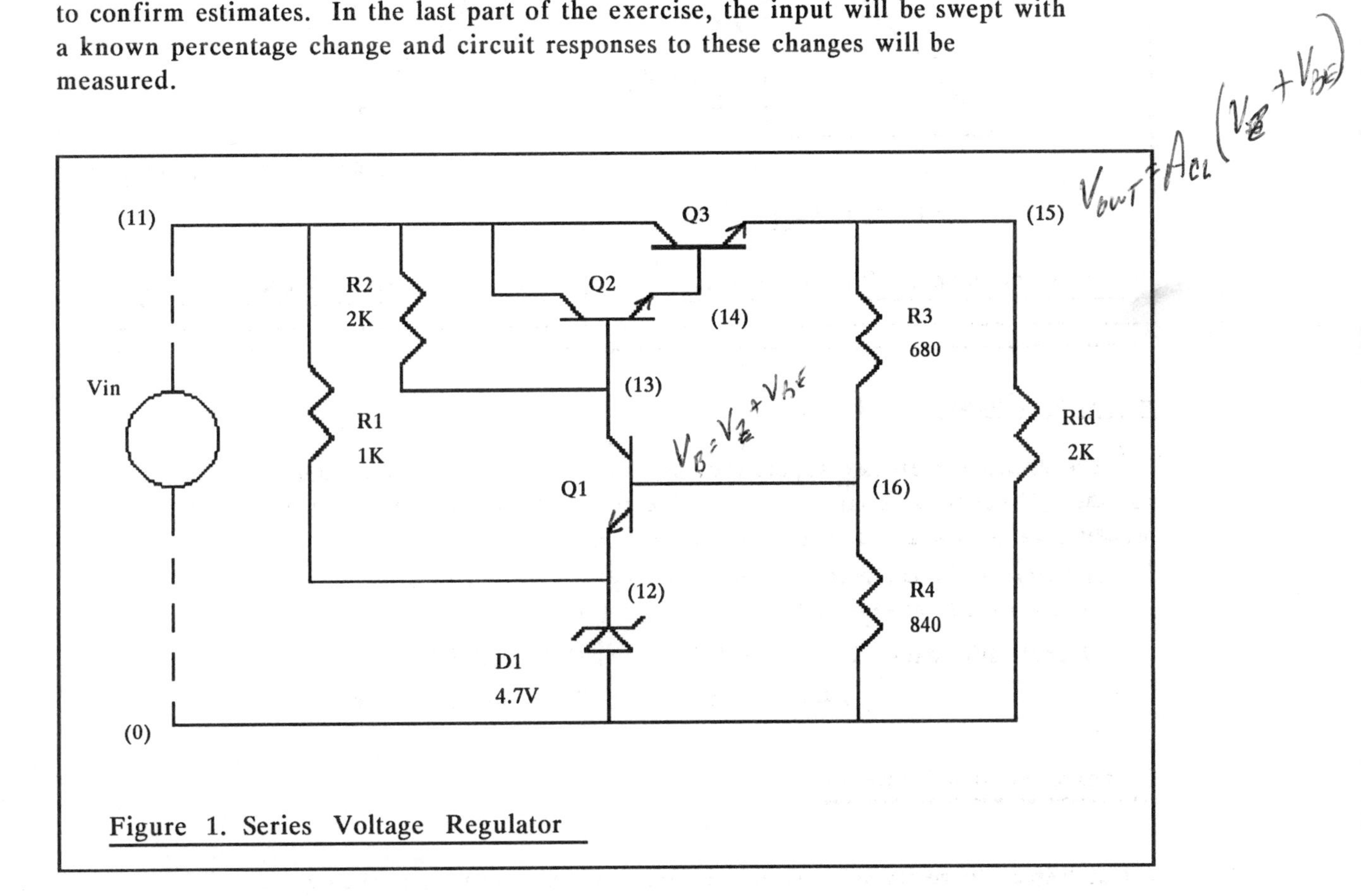

Figure 1. Series Voltage Regulator

Enter the netlist with the Circuit Editor.
Select Files, Enter, Current File, and Enter.
Filename: REGULATOR

```
*SERIES REGULATOR
Vin  11  0  DC  20V
R1   11  12  1K
R2   11  13  2K
D1    0  12  D1N750
Q1   13  16  12  Q2N3904
Q2   11  13  14  Q2N3904
Q3   11  14  15  Q2N2222          ;see netlist  notes
R3   15  16  680Ohms
R4   16   0  840Ohms
Rld  15   0  2K
.Model  Q2N3904  NPN(bf=100 cjc=10 rb=10)
.Model  Q2N2222  NPN(bf=50 cjc=10 rb=8)  ; *
.Model  D1N750  D(Is=880.5E-18 Bv=4.7 Ibv=20m)
.LIB  EVAL.LIB
*.DC Vin 18.6  21.4  .2           ;sweep  input  when  needed
*.Print DC  V(15)
*.Print DC  IC(Q1)  I(R2)  I(D1)
.Probe
.Options Nopage
.End
```

Press Esc to exit and save the circuit file.

Netlist Notes

*The 2N2222 transistor is not a choice for the pass transistor, but is one that is available in the Evaluation Library. The circuit has been optimized for its current and power limits. If a complete PSpice package is available, use suitable power transistor models for Q1, Q2, and Q3.

DC_Sweep

A DC sweep will be inserted to evaluate the operation of the series regulator. The control statement for this input voltage variable is .DC Vin 18.6 21.4 .2. The input to the regulator will be varied 14% by this sweep and each of the points in the .print statements will be examined for percentage change. The Output File and Probe will be used to compare regulator voltages and currents.

DC sweep has been suspended for the first evaluation test. While suspended, the .print statements will be inactive, however, bias points will be calculated and printed in the Output file. Probe also will be inactive without the DC sweep. For the first experiment, the bias points of the circuit will be examined in reference to the regulated 10 Volts at the output.

Circuit Expectations

The Zener voltage, 4.7V, is used to stabilize the series regulator. Zener reference voltage is amplified by Q1, and using negative feedback, provides regulation of the output voltage, V(15), at 10 Volts.

The key to voltage regulation is in the sampling of the output by R3 and R4. This voltage sample is negative feedback to the base of Q1 and controls its amplified collector current. For example, if the output voltage should increase, the voltage divider, R3 and R4, would have a corresponding increase of voltage at the base of Q1. IC(Q1) would increase, providing a larger current through R2 and more V(R2). Less voltage is available at the base of Q2, the first of the Darlington pair pass transistor combination. Less voltage at the base produces more VCE(Q3) and less output at V(Rld). An output voltage decrease would produce the opposite control action and a corresponding increase at the output. The net result is output voltage regulation.

Zener regulation of Q1 emitter voltage at 4.7Volts will place Vz + VBE at the base.

VB(Q1)= 4.7V + .7V = 5.4V (Estimated)

The exact value of Q1 base voltage is needed to calculate closed-loop gain. PSpice will analyze the circuit in a preliminary test and measure this voltage, V(16), for more accurate circuit estimates.

Run PSpice

Select Analysis, Run PSpice, and Enter.
Select Files, Browse Output, and Enter.
Use PgDn, PgUp, and Arrow keys to move within the Output file.
At Small Signal Bias Solutions, record Q1 base voltage, V(16). It should be 5.51 Volts.

V(16), VB(Q1) = ____Volts (Measured)

Closed-Loop Gain Estimates

First Estimate with R4 = 840 Ohms

Closed-loop gain, with R4 = 840 Ohms,

Acl = (R3 / R4) + 1
= (680 / 840) + 1
= 1.81

Output voltage is calculated,

V(15) = Acl(Vz + VBE)
= 1.81(5.51V)
= 10 Volts

Page through the Output file and record the node voltages in Table I. Record source current and power dissipation in the space provided below Table I.

R4 Values	840 Ohms	940 Ohms	740 Ohms
V(11), V(input) →	20v	20v	20v
V(12), V(Zener)	4.699v	4.6997v	4.69v
V(13), VB(Q2)	11.576v	11.10v	12.17v
V(14), VB(Q3)	10.84v	10.36v	11.43v
V(15), V(Output)	10v	9.53v	10.59v
V(16), VF(feedback)	5.5v	5.5v	5.5v

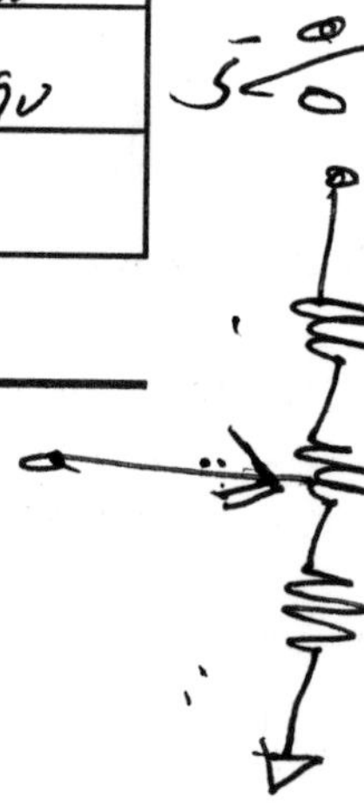

Table I. R4 - Closed-Loop Voltage Variations

Voltage Source Current (R4= 840 Ohms)

I(Vin) = 31.1 mA

Total Power Dissipation = 622 mW

The output voltage, V(15), is near the 10 Volt design level. Adjustment of the output voltage is accomplished by changing the voltage divider network, R3 and R4, thereby changing closed-loop gain.

Adjusting Closed-Loop Gain

Second Estimate with R4 = 940 Ohms

Acl = (680 / 940) + 1
= 1.72

Closed-loop gain is decreased by increasing R4 to 940 Ohms.

Output voltage decreases to,

V(15) = Acl(Vz + VBE)
= 1.72(5.51)
= 9.5 Volts (Estimated)

Edit the Circuit File and Run PSpice

Return to the Circuit Editor and change the value of R4 to 940 Ohms. Press Esc and save the file.
Run PSpice analysis and record the node voltages in Table I.

Third Closed Loop Gain Estimate with R4 = 740 Ohms

Acl = (680 / 740) + 1
= 1.92

Output voltage increases to,

V(15) = Acl(Vz + VBE)
= 1.92(5.51)
= 10.6Volts (Estimated)

Edit the Circuit File and Run PSpice

Return to the Circuit Editor and change the value of R4 to 740 Ohms. Press Esc and save the file.
Run PSpice analysis and record the node voltages in Table I.

Conclusions

We have verified calculations with experimentation. Note each node voltage as the circuit regulates the output to 10 Volts.

Installing A Variable Resistor to Adjust the Output Voltage

Component and device tolerances will require that the finished circuit have some adjustment capabilities. We have the information needed to include this adjustment in the design. From the data in Table I, we see that increasing the size of R4 to 940 ohms decreases the output voltage to approximately 9.5Volts. Also, decreasing the size of R4 to 740 Ohms increases the output to near 10.6 Volts. From these measurements it can be determined that a 150 Ohm variable resistor at V(16) can be used to fine tune the output to the desired 10 Volts. This resistor should be included when constructing the circuit.

Reset R4 to 840 Ohms for 10V output in preparation for the next circuit experiment.

DC Sweep the Input

Unregulated input will be applied to V(11) in this sweep analysis. The series regulator will keep the output, V(15), at near the desired 10 Volt level. Node voltages and circuit currents will be examined for percentage change to illustrate series regulator circuit operation.

Edit the circuit file by removing the asterisk, "*", from the sweep control statement and from the two .print statements. The input will be swept from 18.6V to 21.4V in .2V steps, a swing of 14%. Output file tables will be printed to observe changes in V(15), IC(Q1), I(R2), and I(D1) as the input is swept.

Run PSpice

Select Exit at Probe and return to the Control Shell.

Browse Output

Record high and low measurements in Table II and calculate percentage change.

Table II. Circuit Regulation Values

Node Voltage	Low	High	Percent
Vin	18.6V	21.4V	14%
V(15), V(Output)	9.983	10.018	.35%
Current	Low	High	Percent
IC(Q1)	3.5mA	4.9mA	33.3%
I(R2)	3.5 mA	4.9mA	33.3%
I(D1)	17.5 mA	21.5 mA	20.5%

Sweep percentage change = (High - Low) / Average

= (21.4V - 18.6V) / 20V

= 14%

.DC Sweep Summary

Sweeping the input with a 14% change varies the output voltage less than .4%...less than .04 Volts. Circuit voltages and currents adjust in response to the negative feedback at the base of Q1. Q1 collector current provides regulation to the pass transistor pair.

Using Probe to Observe Circuit Voltage Variations

Select Probe, Run Probe, and press Enter at Control Shell.

V(15), Series Regulator Output

First, set Y_axis range to observe regulator output, V(15):
At Probe; select Y_axis, and Enter.
Select Set_range, and Enter.
At Enter a range: type 9.95V 10.05V , and Enter.
Note the Y_axis range is only .1Volts.
Select Exit, and Enter.

Select Add_trace, F4, arrow to V(15), and Enter twice.
The Probe screen displays the small variations in regulator output. Record total voltage variation: ____V.
Select Remove_trace, Enter, All, and Enter.
Select Exit, and Enter.
Select Y_axis, Enter, Auto_range, Enter, Exit and Enter.

VCE(Q3), Pass Transistor Voltage

Observe VCE(Q3), the pass transistor collector-emitter voltage variations.
Select Add_trace, and Enter.
Type V(11) - V(15) , and Enter.
The screen displays the action of the regulator to compensate for input voltage changes. Record this voltage variation: ____V.

Display the input and output voltages.
Select Add_trace, and display V(11), the input sweep.
Select Add_trace, and display V(15), the regulated output voltage.

As the input, V(11), increases, VCE(Q3) increases, maintaining a near constant 10 Volts at the output, V(15).

Exit Probe

Conclusion of Chapter 15

EOC Problem

(1) Use the series regulator in this chapter and design for a 11 Volt output. Adjust R3 and R4 and calculate the new closed-loop gain. Run PSpice to evaluate your estimates.

Notes:

Chapter 16

THE DIFFERENTIAL AMPLIFIER

Objectives: (1) To use PSpice simulation and analysis of a differential amplifier
(2) To verify differential gain, common-mode gain, and common mode rejection calculations
(3) To practice StmEd editing and Probe "O'scope" waveform analyzing

I. The Differential Amplifier and PSpice

An importance feature of the operational amplifier (op amp) is in the circuit's mathematical operations. Differential amplifier inputs give the op amp its capability to equate mathematically. As the name implies, the output is the amplified difference between the inputs. While amplifying input difference, the differential amplifier has the capability of rejecting, or cancelling, common inputs. Circuit quality is defined using these two quantities.

Simulating and analyzing a differential amplifier begins with defining the netlist in the PSpice circuit file. Figure 1 is the circuit diagram from which the netlist is written. The differential amplifier in this exercise uses two voltage sources, VCC and VEE, and utilizes a current mirror emitter source, Q4. The bias resistor, Rb, is adjusted to provide the required tail current. The mirror current is found with equation,

IC(Q4) = (VCC + VEE - VBE) / Rb = .1mA

This current divides and is the source for Q1 and Q2 emitter currents (.05mA). (Q4 collector current is measured in Problem 1 at the end of the chapter)

The amplifier's differential gain (Adiff), common mode gain (Acm), and common mode rejection ratio (CMRR) will be analyzed in the first section of the chapter. Next, a PSpice exercise is presented for measuring saturation, or overdriving, of the amplifier. In this section, the output waveform will be observed while adjusting input amplitude with the Stimulus Editor. And finally, input equating will be observed by setting the inputs to different frequencies and observing the output with the Probe Graphical Waveform Analyzer.

Examine the circuit on the next page and enter the netlist.

The Test Circuit

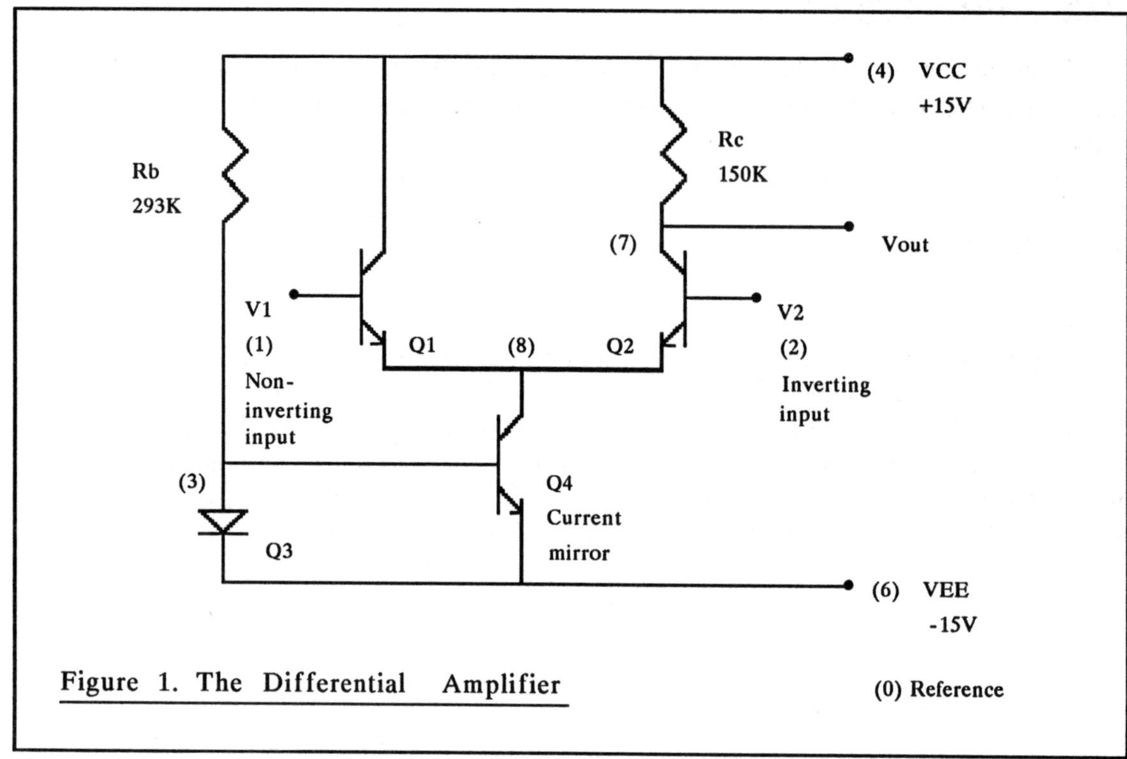

Figure 1. The Differential Amplifier

Enter the netlist with the Circuit Editor.
Filename: DIFFAMP

```
*DIFFAMP - DIFFERENTIAL AMPLIFIER
VCC 4 0 DC 15V
VEE 6 0 DC -15V
Vac1 1 0 AC 1 SIN(0 .02 1KHz)        ;noninverting input
*Vac2 2 0 AC 1 SIN(0 .02 1KHz)       ;inverting input; remove "*" when
*                                    ;statement is needed
Rb 4 3 293K      ;bias resistor
Q3 3 3 6 QNL     ;current mirror diode
Q4 8 3 6 QNL     ;current mirror tx
Q1 4 1 8 QNL     ;noninverting input tx
Q2 7 2 8 QNL     ;inverting input tx
*Rs1 1 0 1K
Rs2 2 0 1K
Rc 4 7 150K
.Model QNL NPN(bf=100 cjc=7p rb=20)
*.AC DEC 5 100m 50Meg
.Tran .02m 2ms 0 .01ms
.LIB EVAL.LIB
.Probe
.Options Nopage
.End
```

Node Voltage Measurements

Suspend the .Tran statement with an asterisk and Run PSpice to measure node bias voltages and source currents. Compare data with differential amplifier theory.

Replace .Tran statement when measurements are completed.

Netlist Notes

Nodes

PSpice requires nodes be connected to at least two points in the circuit. When applying a stimulus to one input of the differential amplifier, it is necessary to connect a resistor from the other input to ground (reference). Rs1 and Rs2 have been included in the netlist for ease of insertion when needed. The "*" makes these two lines comments and removes them from the netlist. Remove the "*" and the resistors are part of the circuit file. Rs2 is connected to Node 2 in the netlist.

Vac1 1 0 AC 1 SIN(0 .02 1KHz)

The first section of this line sets the AC input voltage level for small-signal analysis. AC 1 identifies this input as 1 Volt or unity input. .AC sweep gain measurements are in reference to unity input in this exercise and the output is read as gain (not the unrealistic voltage level). The extremely high input level of 1 Volt is a reference level only and does not disturb gain measurements because PSpice has linearized the circuit and saved the linear equivalent of the circuit for .AC sweep analysis. (See Note on AC Sweep Analysis, page 16-16) Later in the netlist, the .AC sweep small-signal analysis control statement sets sweep frequency limits and the data sample number. (See EOC Problem 2 for gain measurements with other than unity input)

Transient analysis will use the input provided by the stimulus identified in the statement as...SIN(0 .02 1KHz). A sine wave is defined with offset voltage= 0V, amplitude= .02Vp, and frequency= 1KHz. Other stimulus values available are time delay, damping factor, and phase. A default value of "0" is accepted for these undefined quantities.

.AC DEC 5 100m 50Meg

The .AC sweep small-signal analysis statement provides the sweep necessary to measure the frequency response of the amplifier. A log sweep, by decades, with 5 data points per decade sweeping from 100mHz (.1Hz) to 50MHz is needed to display the expected differential amplifier upper frequency limit of approximately 1MHz.

.Tran .02m 2ms 0 .01ms

This transient analysis statement defines print_step= .02ms, final_time= 2ms, results_delay= 0s, and step_ceiling= .01ms.

Print_step is the .Print and .Plot time interval. The analysis begins at 0 time (no delay) and runs until the final time of 2ms, or two hertz of the 1KHz stimulus (input signal). The delay option allows suppressing outputs for a portion of the analysis. Step_ceiling default is final time/50, but is set to .01ms in this analysis. It sets the time of the transient analysis internal time step. The .01ms step_ceiling will provide more evenly drawn Probe graphics.

The transient analysis statement, .Tran, simulates the operation of the circuit in time. If non-linear operation has been introduced into the circuit, either by improperly setting bias points, by overdriving the inputs, or by exceeding the frequency parameters of the circuit, the measurements displayed by Probe will reveal the distortion.

```
.Model  QNL  NPN(bf=100  cjc=7p  rb=20)
```

This line is the model line and calls the library model that will be used in the circuit. A complete definition must be written in the netlist if the model is not available in the library. Parameter values that are different from library model values are written in the netlist when calling for the model. Netlist parameter listings are substituted for library parameters values. In this statement, the QNL NPN bipolar junction transistor model is selected.

The .Model QNL NPN is used in combination with the transistor circuit statements, for example...Q1 4 1 8 QNL. Only three of more than 50 parameters used in setting up the QNL BJT model are defined. They are ideal maximum forward beta, bf=100, base-collector zero-bias p-n capacitance, cjc=7p, and zero-bias (maximum) base resistance, rb=20. The remaining model values are accepted by not listing them. A typical device library listing for a NPN BJT is the 2N2222A model:

```
.Model  Q2N2222A  NPN(Is=14.34f Xti=3 Eq=1.11 Vaf=74.03
+                     Bf=255.9 Ne=1.307 Ise=14.34f Ikf=.2847
+                     Xtb=1.5 Br=6.092 Nc=2 Isc=0 Ikr=0 Rc=1
+                     Cjc=7.306p Mjc=.3416 Vjc=.75 Fc=.5
+                     Cje=22.01p Mje=.377 Vje=.75 Tr=46.91n
+                     Tf=411.1p Itf=.6 Vtf=1.7 Xtf=3 Rb=10)
*                     National       pid=19       case=TO18
*                     88-09-07       bam          creation
```

See Appendix B for parameter definitions.

```
.LIB  EVAL.LIB
```

A control statement to search the Evaluation Library files for the QNL NPN transistor model, or any other model or subcircuit called for in the netlist. The EVAL.LIB entry following .LIB is a must. Many libraries may be included to be used with the PSpice program including a User Library with models defined by anyone accessing the program.

.Probe

A control statement directing PSpice to save all data from the analysis for use by Probe. If no values are specified after .Probe, all node voltages and device currents will be saved in a data file. The data file will have the same name as the circuit file but with a .DAT extension. Extensive data is made available as will be demonstrated later when using the Probe Graphical Waveform Analyzer program.

.Options Nopage

The nopage option will suppress paging and repeating banners in the Output file. Nopage is a paper saving feature when printing Output files. (Print Filename.out).

.End

A necessary statement to signal the end of the circuit file. PSpice simulation and analysis begins only if the circuit file has this statement. Another netlist can be entered after the .End statement and this additional netlist must also conclude with an .End.

A subcircuit, .Subckt, must conclude with an .Ends statement. The 741 op amp used in other chapters is an example of a subcircuit. The UA741 subcircuit is called with an "X" statement in the netlist. It begins with .Subckt and includes all statements following and concludes with .Ends. (See Appendix B)

II. Differential Gain using Transient Analysis

Differential gain is estimated as, Adiff = Rc / 2(r'e), where r'e is ac emitter resistance. AC emitter resistance is calculated using IE. In the test circuit, IE is one-half the tail current or .05mA. From previous studies,

r'e = 25mV / IE = 25mV / .05mA

= 500 Ohms

Adiff = 150K / 2(500)

= 150

Actual gain will be measured later using Probe.

Note the circuit file netlist on page 2. Only one input, Vac1, is needed to measure differential gain. Make sure that Vac2 is removed from the netlist by placing an asterisk at the beginning of the line. Rs2 must be connected at Node 2, the inverting input. PSpice will indicated an error if Rs2 is not connected because this node would have less than two connections. Rs1 and .AC sweep are suspended with asterisks.

The input stimulus at Vac1 is set at .02Vpeak.

Run PSpice

Select Analysis, Enter, Run PSpice, and Enter.

Observe the monitor during Transient analysis. Note the control statement quantities used in the analysis. (.Tran .02m 2ms 0 .01m)

Input Stimulus

The Control Shell program defaults to the Probe Graphical Waveform Analyzer program after a PSpice run. Probe's initial screen should be displayed.

First, take a look at the input stimulus specified in the Vac1 circuit statement. (Vac1 1 0 AC 1 SIN(0 .02 1KHz))

Select Add_trace, and Enter.
Press F4 for a selection of circuit variables.

Note the large number of circuit voltages and currents available to observe. Voltages at each node and currents through each resistor, emitter, base, collector, and voltage source are calculated by PSpice and available to be presented by Probe.

Arrow to V(1) and press Enter twice.

Probe displays the input stimulus at .02Vpeak as defined in the Vac1 statement.

Select Remove_trace, and Enter. Select All, Enter, Exit, and Enter.

Differential Gain Measurements

Next, examine the output voltage at V(7).
Select Add_trace, and Enter. Press F4.
Arrow to V(7) and Enter twice.
The waveform has a peak-to-peak amplitude of approximately 5.5V. Vpeak= 2.75V.

Actual Differential Gain,

Adiff = 2.75 Vpeak / .02 Vpeak
= 137.5

Differential gain at 137.5 is within 10% of estimated value (150).

Repeat this measurement with Vac1 inputs of .01V, .005V, .001V, and .0001V. Differential gain should be approximately 140 for each test.

Use the Circuit Editor to edit input values.

Return to Control Shell after completing differential gain measurements.

III. Differential Gain using AC Sweep Analysis

Edit the circuit file for AC_sweep analysis. Remove the "*" and place the AC sweep analysis control statement into the netlist. Set Vac1 to .02Volts input.

```
*DIFFAMP - DIFFERENTIAL AMPLIFIER
VCC  4  0  DC  15V
VEE  6  0  DC  -15V
Vac1  1  0  AC  1  SIN(0  .02  1KHz)
*Vac2  2  0  AC  1  SIN(0  .02  1KHz)
Rb   4 3 293K      ;bias resistor
Q3  3  3  6  QNL   ;current  mirror  diode
Q4  8  3  6  QNL   ;current  mirror  tx
Q1  4  1  8  QNL   ;noninverting  input  tx
Q2  7  2  8  QNL   ;inverting  input  tx
*Rs1  1  0  1K
Rs2  2  0  1K
Rc  4  7  150K
.Model  QNL  NPN(bf=100 cjc=7p rb=20)
.AC  DEC  5  100m  50Meg
.Tran  .02m  2ms  0  .01ms
.LIB  EVAL.LIB
.Probe
.Options Nopage
.End
```

The .AC sweep statement will calculate a small-signal frequency response of the differential amplifier. .AC DEC 5 100m 50Meg will provide a log sweep, by decades, with 5 points per decade. Vac1 input will be swept from .1Hz to 50MHz.

Run PSpice

AC_Sweep Measurements

The Probe Graphical Waveform Analyzer title screen should be displayed.
Arrow to AC_sweep, and Enter, to measure amplifier response to the input sweep.
Select Add_trace, and Enter. Press F4 for a list of variables available for viewing. All circuit voltages and currents are presented.
Arrow to V(7), the output at Q2 collector, and Enter twice.

The graph displays output gain (not volts) in reference to unity input (AC 1 in the Vac1 statement). Gain is approximately 138. Critical frequency gain is near 100 and upper critical frequency is below 1MHz.

Select Remove_trace, and clear the screen for decibel gain measurement. (Decibel gain = (20)(log Adiff= 137.5) = 42.77dB)

Select Add_trace, and Enter.

Type Vdb(7), and Enter.

Decibel gain is measured at approximately 43dB.

Next, take a look at circuit input quantities.

Select Remove_trace, and clear the screen.

Select Add_trace, and display V(1).

Input sweep is displayed as a steady 1Volt from .1Hz to 50MHz.

Select Exit, and Enter, to return to Probe title screen.

Transient Analysis Measurements

Select Transient_analysis, and Enter.

Select Add_trace, and display V(7), the output.

The sine wave measures 2.75Vpeak. Gain= 137.5 as measured previously. (2.75Vp/.02Vpeak= 137.5)

Select Remove_trace, and clear the screen.

Additional Differential Amplifier Measurements

Take a look at other amplifier voltages and currents available in Probe options. Study each waveform display in reference to differential amplifier theory. Individually view V(1), V(2), and V(8).

Simultaneously view IE(Q1) and IE(Q2), the input transistors' emitter currents. Note the .05mA emitter bias current level.

Select Exit, and Enter, to leave Probe menu screen.

Exit Probe.

IV. Common Mode Gain Measurements using Transient Analysis

Common mode gain, Acm, is estimated as Acm = Rc/2RE. In this circuit, the emitter resistance is the current mirror source with an extremely high impedance, in the order of megohms. Therefore, Acm will be extremely low, in the order of .02 assuming a 5-10Meg emitter resistance.

Using the Circuit Editor, edit the circuit file as presented on the next page. Note the inputs at Vac1 and Vac2. It will be necessary to analyze the circuit with different input amplitudes to verify Acm. Stimuli amplitudes will be adjusted for each PSpice run; .2Vpeak, .4Vpeak, .8Vpeak, and 1Vpeak. In each instance, the output will be measured and divided by peak input to obtain Acm.

Note the high levels of input compared with .02Vpeak and less used in differential gain measurements. Common mode rejection testing in the differential amplifier requires the higher inputs to increase the output to acceptable measurement levels.

In the first test, Vac1 and Vac2 are providing the same common-mode stimuli to the inputs and are set to .2Vpeak.

StmEd will be used to adjust the inputs during the exercise. Repeated use of StmEd will improve efficiency in the program.

Circuit file for common mode measurements.

```
*DIFFAMP - DIFFERENTIAL AMPLIFIER
VCC  4  0  DC  15V
VEE  6  0  DC  -15V
Vac1  11 0 SIN(0 .2 1KHz)       ;Vpeak= .2V for transient
Vac2  12  0  SIN(0  .2  1KHz)  ;analysis,  edit to .4V, .8V,
*                              ;and 1Vpeak  later  in exercise
R10  11  1  100K     ;resistor  at non-inverting  input
R11  12  2  100K     ;resistor  at inverting  input
Rb   4  3  293K      ;bias  resistor
Q3  3  3  6  QNL     ;current  mirror  diode
Q4  8  3  6  QNL     ;current  mirror  tx
Q1  4  1  8  QNL     ;noninverting  input  tx
Q2  7  2  8  QNL     ;inverting  input  tx
Rc  4  7  150K
*Rs1  1  0  1K       ;suspended  for common  mode  measurements
*Rs2  2  0  1K       ;suspended  for common  mode  measurements
.Model  QNL  NPN(bf=100 cjc=7p rb=20)
*.AC  DEC  5  100m  50Meg
.TRAN  .02m  2ms  0  .01ms
.LIB  EVAL.LIB
.Options Nopage
.Probe
.End
```

Press Esc to exit and save the file.
Sketch the new input stimuli circuits.

Stimulus Editor

Browse the Stimulus Editor program and take a look at the two differential amplifier inputs.

Select StmEd, Enter, Edit, and Enter. Both inputs are displayed superimposed on the initial screen.

The amplitude for Vac1 and Vac2 should be .2Volts.
Select Modify_stimulus, and Enter.
Press spacebar to select Vac1 to be modified, and Enter.

This StmEd screen displays a sine wave and the Vac1 stimulus parameters:

Transient Spec Type	: SIN
VOFF (offset voltage)	: 0
VAMPL (amplitude)	: .2
FREQ (frequency)	: 1.000E3
TD (time delay)	: 0
DF (damping factor)	: 0
Phase (phase shift)	: 0

Vac1 stimulus parameters are; offset voltage= 0, amplitude= .2Vpeak, damping= 0, frequency= 1KHz, delay= 0, and phase= 0. Vac2 is providing the same stimulus values for common-mode measurements.

We will return to edit these inputs later in this exercise. For now, leave StmEd, and run PSpice analysis of the differential circuit to measure Acm.

Select Exit, and Enter, to leave Modify_stimulus screen.
Select Exit, and Enter, to leave StmEd menu screen.
Select Exit_program, and Enter, to leave the Stimulus Editor program.

Run PSpice For Common Mode Gain Measurements, Acm

At Control Shell

Select Analysis, Enter, Run PSpice, and Enter.
PSpice performs a transient analysis on the circuit as specified in the .Tran statement.

At Probe

Select Add_trace, and display V(7).

That rather noisy output is the small signal surviving common mode inputs. Record estimated Vpeak= _______V.
(V(7) should be approximately .004Vpeak)

Using output / input, calculate Acm= _______. (Input= .2Vpeak)
(Common Mode Gain, Acm, should be approximately .02)

Select Exit as needed to leave Probe and return to Control Shell.

StmEd Editing

Using StmEd, edit Vac1 and Vac2 to .4Vpeak.

Select StmEd, Enter, Edit, and Enter as needed to access the Stimulus Editor.
Select Modify_stimulus, and Enter.
Arrow to either stimulus to modify, press Spacebar to select, and Enter.
Select Transient_parameters, and Enter.
Arrow to 2)VAMPL, and Enter.
Backspace to erase, then type .4V, and Enter.

Select Exit, and Enter, to leave Transient_parameter screen.
Select Exit, and Enter, to leave Modify_stimulus screen.
Repeat the above procedure and modify the other stimulus.

Exit StmEd when completed.

Note: Use the Circuit Editor and take a look at the modified stimuli after editing the netlist with StmEd.

Run PSpice

At Probe

Record V(7)= _______Vpeak. Calculate Acm= _______. (Input= .4Vpeak)

Return to Control Shell and repeat the StmEd procedure listed above to edit Vac1 and Vac2 to .8Vpeak.

Run PSpice

Record V(7)= ______Vpeak. Calculate Acm= _______. (Input= .8Vpeak)

Use StmEd to edit Vac1 and Vac2 to 1Vpeak.

Run PSpice

Record V(7)= ______Vpeak. Calculate Acm= _______. (Input= 1Vpeak)

Common Mode Gain, Acm, should be approximately .02 - .04 in these exercises.
This concludes experimenting with Acm measurements.

V. Common Mode Rejection Ratio

Common mode rejection ratio is estimated as,

CMRR = Adiff / Acm

= 150 / .02 (Estimated)

= 7500

Actual,

CMRR = Adiff / Acm

= 137.5 / .02

= 6875

VI. Saturation Measurements using Transient Analysis

(An Exercise)

The following measurements overdriving the amplifier are entered for the purpose of gaining experience using PSpice programs. Increasing differential input causes waveform distortion in the amplifier output. At sufficiently high input levels the circuit saturates and waveform clipping occurs.

The current mirror in the emitter circuit locks emitter current to .05mA in each input transistor. The voltage across Rc is approximately 7.5V and the output collector quiescent voltage is set to near 7.5Volts. With differential gain at 137.5, and a maximum output swing of 7.5Volts, the input to cause distortion can be estimated at,

Vin(sat) = 7.5V / 137.5
= .055Vpeak

Using the Circuit Editor, edit the circuit file in preparation for distortion level testing. Vac1 will be set to .06Vpeak in the first measurement.

```
Vac 1 0 AC 1
+ SIN( 0 .06 1.000E3 0 0 0 )
*Vac2  2  0  AC  1
* + SIN( 0 .06  1.000E3  0  0  0 )
*R10  11  1  100K
*R11  12  2  100K
*Rs1  1  0  1K
Rs2  2  0  1K
*.AC  DEC  5  100m  50Meg
```

All other circuit description and control lines remain unchanged.

Run PSpice

Probe's initial screen should be displayed.
Select Add_trace, and Enter.
Press F4 and arrow to V(7), the output node.

Non-linear distortion is occurring, but no clipping that would indicate complete saturation of the differential amplifier. The output is being driven from 1.8V to 13.8V, a 12Vpp swing. Adiff has dropped to 100 as a result of the high input and distortion. (6Vpeak(out) / .06Vpeak(in)= 100)

Run PSpice with inputs of .08Vpeak, .09Vpeak, .1Vpeak, and .3Vpeak. Notice output waveform distortion at each input level and recalculate gain. Clipping is occurring above .08Vpeak input.

Note: A stimulus of .02Vpeak seems to be the maximum input without some output distortion. (Adiff= 137.5) At input= .03V, Adiff= 126. At input=.04Vpeak, distortion is increased and Adiff= 117.

Change the input to Vac2, the inverting input, and run the same test. To edit, change only the following lines:

```
*Vac1 1 0 AC 1
* + SIN( 0 .06 1.000E3 0 0 0 )
Vac2 2 0 AC 1
+ SIN (0 .06 1.000E3 0 0 0 )
Rs1 1 0 1K
*Rs2 2 0 1K
```

Run PSpice

With the input at Vac2, the only change should be the output is now inverted.
This concludes saturation measurements.

VII. Differential Amplifier Input Equating

(An Exercise)
In this part of the exercise, different amplitudes and frequencies will be applied to the diff amp inputs. Input differences will be amplified by the circuit. StmEd will be used to change the amplitude and frequency at the inputs. Probe will plot the output, V(7).

Edit Netlist

The netlist on page 16-9 will be used in this exercise. Edit this netlist into the circuit file. Next, use StmEd to set the input stimuli.

StmEd Editing

Select StmEd, Enter, and follow the procedure to edit Vac1 to .01Vpeak.
Select Exit, and Enter, to leave Transient_parameters screen.
Repeat for Vac2, setting this input to .01Vpeak.

Next, set Vac2 frequency to 3KHz.
Select Modify_stimulus, and Enter.
Arrow to Vac2, and press Spacebar to select Vac2 to be modified. Press Enter.
Select Transient_parameters, and Enter.
Select 3)FREQ, and Enter.
Backspace and erase 1.0E3, and type 3000, and Enter.
Select Exit, and Enter, to leave Transient_parameters screen.
Select Exit, and Enter, to leave Modify_stimulus screen.

The two waveforms on the StmEd screen are equal in amplitude. Vac1 has a frequency of 1KHz and Vac2 is set to 3KHz.

Select Exit, and Enter, to leave StmEd menu screen.
Select Exit_program, and Enter, to leave StmEd. and return to the Control Shell.

Examine these edited inputs with the Circuit Editor before running PSpice analysis.

Run PSpice

At Probe

Select Add_trace, and Enter.
Press F4 for a selection of variable to plot.
Arrow to V(7), and Enter twice.

The output waveform, V(7), is the amplified amplitude difference between the two input stimuli that have frequencies of 1KHz and 3KHz. This display can be compared to the familiar two-tone waveform in telephone tone dialing systems.

Exercises

(1) Repeat the above procedure and edit Vac2 to 10KHz. Run PSpice and view with Probe. Compare the output waveform to the output in the previous exercise (Vac2 set to 3KHz).

(2) Edit Vac1: frequency=100KHz, amplitude=.01Vpeak.
Edit Vac2: frequency=1KHz, amplitude=.01Vpeak.
Run PSpice and Probe and view the output, V(7).
The 100KHz waveform appears superimposed on the 1KHz signal. The differential amplifier output is the amplified amplitude difference between the inputs.

(3) Repeat Exercise 2 and set Vac1 to .3mV at 1KHz, and Vac2 to .01V at 100KHz. The 1KHz stimulus appears superimposed on the 100KHz waveform.

Conclusion of Chapter 16

EOC Problems

(1) Complete the following measurements to compare current mirror current, IC(Q4), with bias resistor current, I(Rb). Enter the netlist as listed on page 16-2 but suspend .AC sweep and .Tran analysis. Insert the following DC sweep and print lines:

```
.DC  VCC  15  15  1
.Print  DC  I(Rb)  IC(Q4)  IE(Q1)  IE(Q2)
```

Run PSpice and Browse Output.

Collector current (Q4) should be approximately the same as bias resistor current. Q1 and Q2 emitter currents should be one-half Q4 collector current.

(2) Complete the following exercise to measure AC sweep differential voltage and decibel gain with an input level other than unity. Using the netlist on page 16-2, edit the following changes:

Set AC 1 in the Vac1 statement to AC .001, for 1mV input.
Remove the asterisk and insert the .AC sweep control line.
Make the .Tran line a comment.

Run PSpice
Select Add_trace at Probe and display V(7), the output.

Output is measured at approximately 138mV, and differential gain is calculated as,

Adiff= output / input = V(7) / V(1)

= 138mV / 1mV

= 138

Differential voltage gain is the same as with unity input used earlier in the chapter.

Select Remove_trace and clear the screen for decibel gain mesurements.

Select Add_trace, and Enter.

Since the input is not unity (1V), PSpice must be programmed to calculate decibel output/input measurements.

Type db(V(7)/V(1)), and Enter.

Gain is approximately 43dB, the same as previously measured.

A Note on AC Sweep Analysis

The input levels, AC 1 or AC .001, are insignificant since the circuit has been linearized for AC Sweep measurements. (Linearized means the AC Sweep analysis is performed on the linear component equivalent of the circuit) The output amplitude at V(7), or any other node amplitude, is relative to (in proportion to) the input at Node 1. Measurement results are valid where frequency limits are not exceeded and circuits are not driven into nonlinear operation.

(3) Select a differential amplifier from a theory textbook and, after completing all calculations, write the netlist and Run PSpice to verify you estimates.

Chapter 17

LINEAR OPERATIONAL AMPLIFIER

Objectives: (1) To use PSpice to check critical frequencies in a 741C Op Amp circuit
(2) To use Probe Graphical Waveform Analyzer AC_sweep and Transient_analysis to display 741C Op Amp behavior

A Linear 741 Op Amp Circuit

PSpice simulation and analysis of the differential amplifier has been examined in a previous chapter. The differential amplifier is the input stage to the operational amplifier in the present exercise. The 741C integrated circuit will be used as a bandpass amplifier. Circuit components will limit the lower end of the spectrum to a critical frequency of approximately 160 Hz, and the upper end to approximately 10 KHz. Critical frequency component design will be included in the next chapter.

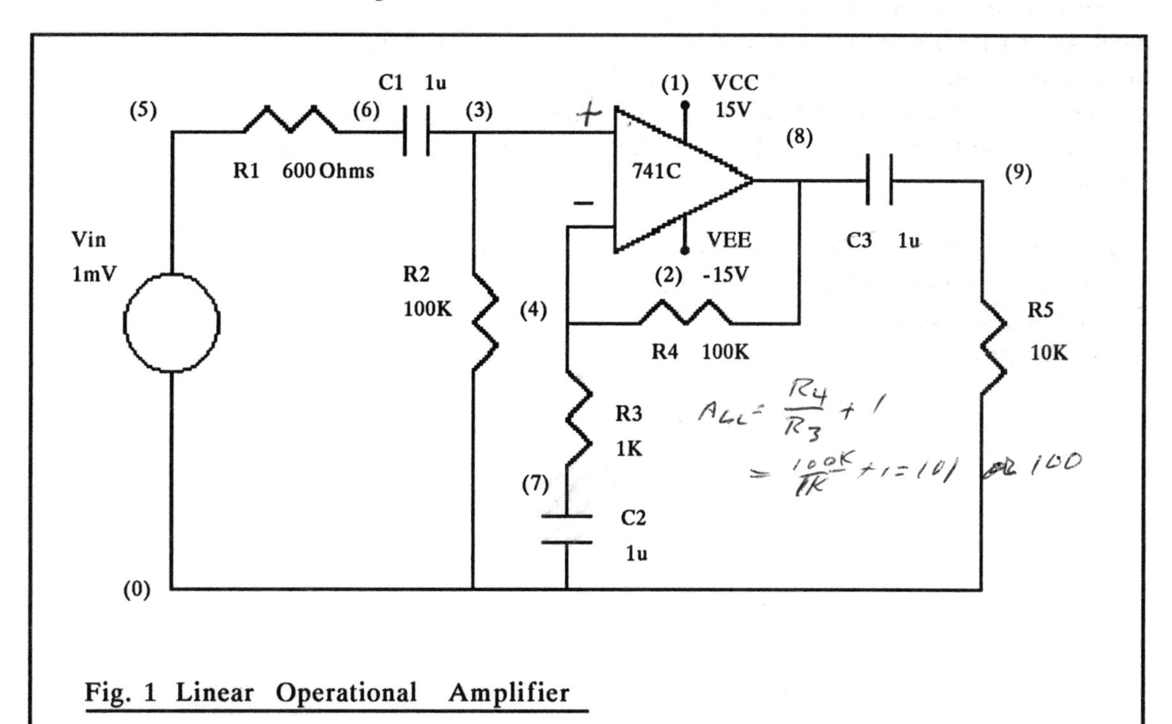

Fig. 1 Linear Operational Amplifier

The Circuit File

Type PS at the PSEVAL52 directory.
Select Files, Enter, Current File, and Enter.
Filename: LINOPAMP
Select Edit, and Enter.

Enter the netlist with the Circuit Editor.

```
*LINOPAMP - LINEAR OP AMP
VCC 1 0 DC 15V      ;positive power supply
VEE 2 0 DC -15V     ;negative power supply
Vin 5 0 AC .001V SIN(0 0.001V 5KHz) ;input stimulus
R1 5 6 600Ohms
C1 6 3 1u
R2 3 0 100K
R3 4 7 1K
R4 4 8 100K
C2 7 0 1u
C3 8 9 1u
R5 9 0 10K                  ;load resistance
XAMP 3 4 1 2 8 UA741
.LIB EVAL.LIB
.AC DEC 5 10m 100K
.Tran .004ms .4ms 0 .002ms
.Probe
.Options Nopage
.End
```

Press Esc to save the circuit file.

Netlist Notes

Vin 5 0 AC .001V SIN(0 0.001V 5KHz)

This circuit netlist statement sets the input stimulus specifications for AC_sweep analysis and defines the circuit stimulus for transient_analysis.

AC .001V identifies the small-signal analysis input level. Later in the circuit file, the .AC control statement places sweep limits on the input as well as defines the number of data points, 5 per decade in this analysis.

SIN(0 0.001V 5KHz) sets the input stimulus for transient analysis. Entries in parenthesis are offset_voltage= 0V, amplitude=.001V, and stimulus frequency= 5KHz. Three other entries are omitted and default values are accepted. They are time_delay= 0 seconds, damping_factor= 0, and phase= 0 degrees.

The .Tran statement is listed later in the netlist and defines the required analysis of the 5KHz signal. Transient analysis stimulus parameters can be entered or changed with the Circuit Editor or with the Stimulus Editor program.

.AC DEC 5 10m 100K

The AC_sweep control statement provides the sweep analysis necessary to measure the frequency response of the amplifier. A log sweep with 5 data points per decade sweeping from 10mHz (.01 Hz) to 100KHz is set to cover the expected bandwidth. The bandwidth of this amplifier is expected to be 160Hz - 10KHz.

The .AC statement will sweep the circuit input, Vin, with the frequencies specified. Probe's sweep analysis program will display the output as a representation of the bandwidth of the circuit. It is important to keep in mind that the frequency and voltage input limits of the circuit will not be exceeded and Class A linear operation can be expected.

.TRAN .004ms .4ms 0 .002ms

The first entry in the transient analysis line is .004ms. It is the print_step time interval for .print and .plot results from transient analysis. This time is also stored for the Probe Graphical Waveform Analyzer program. Print_step time is calculated as 1/100 of the final_time for this analysis. (.4ms/100=.004ms)

Final_time is the next entry and is .4ms. It has been set to display two hertz of the 5KHz input. Final_time is calculated as t= 2(1/f) and is the actual time for transient analysis to run.

The third entry is results_delay and is set to 0 time. It suppresses the transient analysis start for zero time (no delay).

Step_ceiling is set to .002ms and is the last entry in the .Tran statement. Step_ceiling default time is calculated as final_time/50, .4ms/50= .008ms, and is the maximum step_ceiling time for transient analysis. In this analysis, final_time/200 is used for better Probe graphical displays. (.4ms/200= .002ms)

The .Tran statement simulates the operation of the circuit in time, specifically during 2 hertz of the 5KHz input stimulus (input signal). If non-linear operation has been introduced into the circuit, either by setting the bias points, by overdriving the inputs, or by exceeding the frequency limitations of the circuit, the outputs that are examined by Probe will reveal the non-linearity. The circuit is designed for an upper critical frequency of approximately 10KHz, therefore, the 741C's unity gain frequency of 1MHz is not approached. The noninverting input of 1mVpeak will be within specs, therefore, the transient analysis display of circuit output, V(9), should not indicate waveform distortion or clipping.

XAMP 3 4 1 2 8 UA741

This is the "X" statement that calls for the UA741 op amp subcircuit used in the netlist. The UA741A op amp subcircuit is included in the Evaluation Library and definition data is drawn from this library. The subcircuit listing is very similar to a macro line. Numbers define the UA741 connections by position versus use.

Position	Node No.	Use
First position	3	Noninverting input connection
Second position	4	Inverting input
Third position	1	Positive voltage source
Fourth position	2	Negative voltage source
Fifth position	8	Output

Position No. : 3 1 741C 8 4 2

Fig. 2. Op Amp Subcircuit Node Listings

Note that the node numbers used in the netlist are not necessarily the position numbers. A designated node connection on the op amp must correspond with the position in the library subcircuit. Any node number may be used if that number is placed in the proper position.

.LIB EVAL.LIB

The control statement directing the PSpice program to search the Evaluation Library for the XAMP UA741 subcircuit. The EVAL.LIB entry is necessary since many libraries may be included to be used with PSpice.

.Options Nopage will eliminate paging and repeating banners in the Output file.

.End indicates the end of the circuit. PSpice simulation and analysis begins only after the .End statement. Another circuit netlist can be entered after .End and the additional circuit must also include an .End.

PSpice Simulation and Analysis

Run PSpice

Note: It is not necessary to run PSpice again to return to the Probe program. Probe can be used at any time to observe data files (.DAT) by going directly to Probe in the Control Shell or at the DOS prompt by typing Probe, and pressing Enter. It is necessary to Run PSpice only if the circuit file has been changed.

AC_Sweep Analysis

Select AC_sweep at Probe title screen, and Enter.
Select Add_trace, and Enter.
Press F4 for a list of variables and expressions available.
Select V(9) for the output at the load resistor, R5, and Enter twice.

The Probe display is the familiar bell-shaped curve of a bandpass amplifier bordered by lower and upper critical frequencies. The lower critical frequency is approximately 160 Hz and the upper frequency is near the expected 10 KHz, as expected. But note the upper frequency response is cut off by the upper sweep limit (100KHz) that was placed in the AC sweep statement.

Return to the Circuit Editor and adjust sweep parameters to present a more complete amplifier bandwidth response.
Select Exits, and Enter, and return to Control Shell.
Select Files, Enter, Edit, and Enter.
Edit .AC DEC 5 10m 100K to .AC DEC 5 100m 200K.
The sweep is changed from .01Hz - 100KHz to .1Hz - 200KHz

Run PSpice

AC_Sweep Gain Measurements

Select AC_sweep, and Enter.
Select Add_trace and display the output, V(9). The upper end of the sweep is extended to cover the upper end of amplifier bandpass.
Gain, at critical frequencies (160Hz and 10KHz), can be estimated at 70.7% of maximum, approximately 70. (Maximum gain= 100mV/1mV= 100)
Select Remove_trace, and clear the screen.
Select Add_trace, and type db(V(9)/V(3)), and Enter to measure decibel gain.
Set Y_axis range to 0 - 40 to improve the display.
Gain is approximately 37dB at critical frequencies.
(Gain(dB) = (20)(log 70) = 36.9dB)
Select Auto_range, and Enter, to reset Y_axis. Select Exit, and Enter.
Select Remove_trace, and clear the screen.

Input Sweep

Select Add_trace, and display V(3), the sweep input.
Sweep input is 1mV over the range specified in .AC sweep.
Select Exit, and Enter, to return to Probe title screen.

Transient_Analysis

In transient analysis, Probe displays the 5KHz waveform for examination. If the frequency limits of the op amp circuit have not been exceeded, and if the amplifier has not been overdriven, we should expect a clean sine wave output at V(9). The circuit is within these limits.

Select Transient_analysis, and Enter.

The transient analysis screen is displayed with Time on the X_axis, 0s - 400us.

Input Stimulus

Select Add_trace, and Enter.
Press F4 and arrow to V(5), the input stimulus, and Enter twice.

Two hertz of the 5KHz input signal are displayed. Amplitude is measured at 1mVpeak.

Examine V(6) for the same signal.
Select Remove_trace, and clear the screen.
Select Add_trace, and display V(3), the non-inverting input of the op amp.
Two hertz at 1mVpeak is displayed, with an offset voltage of -8.0mV.

We will take another look at the input stimulus with the StmEd program at the close of this exercise.

Select Remove_trace, and clear the screen.

The Output Signal

Select Add_trace, and display V(9), the output node voltage.

Transient analysis was set to run for two hertz at 5KHz (.4ms). This frequency is off the center frequency of the amplifier bandpass, therefore, gain is less than maximum (100). The Probe screen displays two undistorted hertz with gain of approximately 90.

(vp(out) / vp(in)= 90mV / 1mV = 90)

Select Remove_trace, and clear the screen.

Display Input and Output Waveforms Simultaneously

Select Add_trace, and display V(3), the non-inverting node.
The input stimulus is presented, 1mV at 5KHz.

Select Add_trace, and display the output node, V(9). The input signal is very small compared to the output. As expected, the output is in-phase with the non-inverting input.

Experiment with changing the Y_axis range to improve the display of the input stimulus. (-10mV 10mV)

Select Exit, and Enter, to return to Probe title screen.

Select Exit_program, and Enter, to return to the Control Shell.

A short overview of the input stimulus with StmEd and a look at the Output file will complete this chapter analyzing the operation of a linear 741C op amp circuit.

Stimulus Editor

Select StmEd, Enter, Edit, and Enter.

The StmEd screen displays 2 hertz of the 5KHz input stimulus at 1mVpeak.

Select Modify_stimulus, and Enter, to review the Transient_parameter spec table.

Select Transient_parameters, and Enter.

Step through the parameter adjustments available in StmEd by pressing Enter. When finished, select Exit, and Enter to leave the Transient_parameter menu screen.

Exit Stimulus Editor.

Browse Output

Select Files, and Enter.
Select Browse Output, and Enter.
Browse and evaluate the Output file.

Screens listed are:

(1) Circuit Description
(2) Diode Model Parameters
(3) BJT Model Parameters
(4) Small Signal Bias Solution
(5) Initial Transient Solution

Total Power Dissipation
Job Concluded
Total Job Time

XAMP node connections listed in the Output file are part of the UA741 subcircuit netlist and can be examined in Appendix B.

Question: How did screens (2) and (3) get into the Output file?

Ans: The Diode Model Parameters and BJT Model Parameters listed in the Output file are part of the UA741 subcircuit library listing. Take a look at the UA741 subcircuit model in the EVAL.LIB and in Appendix B.

.Models within the subcircuit are not available outside the subcircuit and may not be called in the netlist. However, .models in the netlist may be used by subcircuits.

Question: How is the PSpice user able to arbitrarily assign node numbers without consideration for node numbers designated in subcircuits?

Ans: Node labels in subcircuits are unique to the particular subcircuit. There is no chance of duplication. Run PSpice and examine the Output file in this chapter for examples of XAMP subcircuit label assignments. Examples are (XAMP.6) and (XAMP.7).

Press Esc to exit Output Browser.

Exit PSpice Control Shell.

Conclusion of Chapter 17

EOC Problem

(1) Select several 741C amplifier circuits from theory textbooks and lab manuals and Run PSpice simulation and analysis to verify designs.

(2) Review the op amp circuit and experimentation in this chapter and write a two page paper comparing hardware breadboard testing with PSpice simulation and analysis.

Chapter 18

LINEAR BANDWIDTH AMPLIFIER

Objectives: (1) To analyze the design of circuit components used to select bandwidth
(2) To use PSpice and Probe presentations to analyze bandwidth
(3) To examine the results of overdriving op amp input
(4) To use StmEd to modify the input stimulus

Component Design for Linear Op Amp Circuit Bandwidth

In Chapter 17, we used PSpice and Probe to observe the bandpass of a 741 op amp circuit. In this chapter, the design of components that determine upper and lower critical frequencies in a linear amplifier is considered. Op Amps typically have very high gain at frequencies from DC to >1MHz. Both gain and bandwidth are set with resistors and capacitors and a feedback network. The circuit below will be used to evaluate the design of these components.

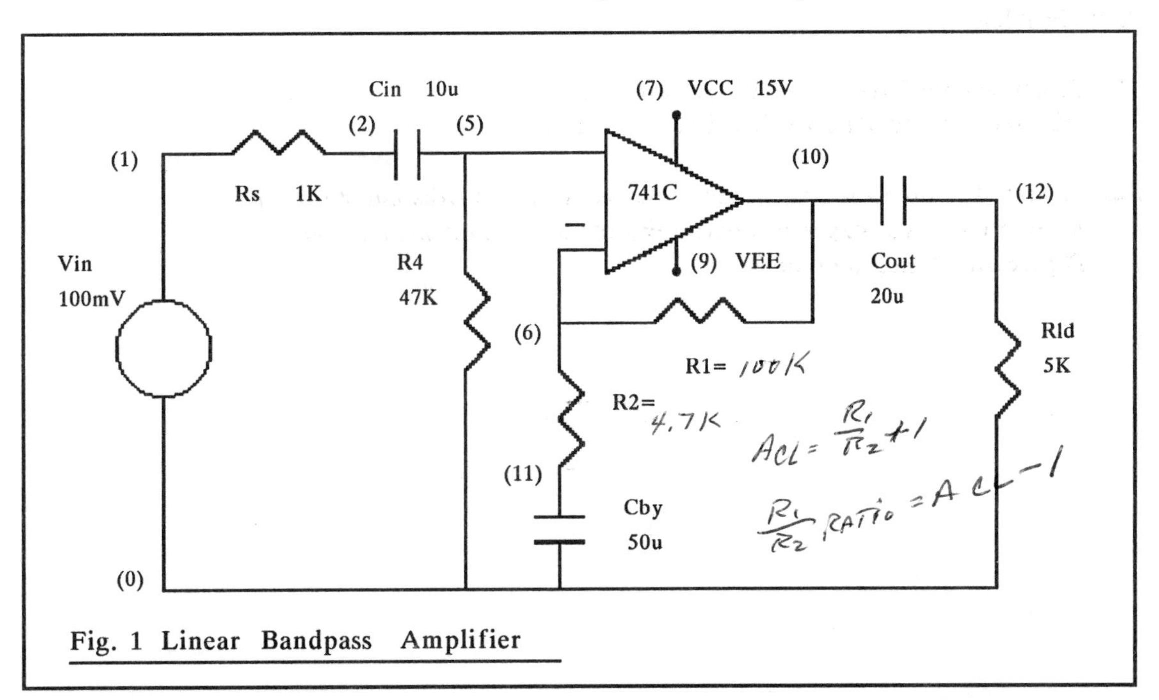

Fig. 1 Linear Bandpass Amplifier

UA741 Library Subcircuit

The UA741 subcircuit is connected with a different set of node numbers to illustrate the position vs use relationships of node selection. (See Appendix B) Numbers assigned to nodes are, of course, insignificant. But when listing the XAMP call in the netlist, it is a must to observe circuit node number vs subcircuit connection position. Review the following XAMP statement.

XAMP 5 6 7 9 10 UA741 This line is the call for the op amp subcircuit. The subcircuit listing in the library is very similar to a macro line. Node numbers define the UA741 connections in the circuit. Observe subcircuit node vs position use, and nodes assigned:

First position	5	Noninverting input	Position/Node No.
Second position	6	Inverting input	5 7
Third position	7	Positive voltage source	10
Fourth position	9	Negative voltage source	6 9
Fifth position	10	Output	

The Feedback Divider and Gain

The feedback divider circuit, R1 and R2, will help to define gain in this non-inverting voltage feedback amplifier. From previous studies we have seen closed-loop voltage gain as,

Acl = (R1 / R2) + 1 (Equation 1)

We are designing for a gain/bandwidth tradeoff to provide a bandwidth, lower critical frequency (F1) to upper critical frequency (F2), of approximately 45KHz. Assuming F1 to be near DC, and a unity gain frequency of 1MHz for the 741 op amp, then gain can be estimated,

Acl = F(unity) / F(upper critical frequency) (Equation 2)

Acl = 1MHz / 45KHz = 22.2

The R1 / R2 ratio in Equation 1 = (Acl - 1) = (22.2 - 1) = 21.2. Assuming an R1 value of 100K, and rearranging Equation 1 to solve for R2,

R2 = R1 / (Acl-1) = 100K / (22.2 - 1) = 4.7KOhms

Enter these values for R1 and R2 in the circuit diagram and in the netlist on page 18-4.

Note that decreasing closed-loop gain from the value in the previous chapter has significantly increased the upper critical frequency, therefore, amplifier bandwidth has been increased.

Lower Critical Frequency Considerations

Since the lower end of the bandpass is limited by the highest (or dominant) lower critical frequency (F1) of the circuit, the critical frequencies of the input, bypass, and output circuits must be calculated. In each section,

F(critical) = 1 / (6.28)(Rseries)(C)

The input coupling capacitor, Cin= 10ufd, has a total series resistance of Rs + R4 = 48KOhms. Input critical frequency is,

Fc(input) = 1 / (6.28)(48E3)(10E-6)
= .3Hz

The series resistance of the bypass capacitor, Cby= 50ufd, is the value of R2= 4.7KOhms. The inverting input is a virtual ground.

Fc(bypass) = 1 / (6.28)(4.7E3)(50E-6)
= .7Hz

Output coupling has a 5KOhm series load resistance with C3= 20ufd. Output impedance of the closed-loop amplifier is approximately 0 Ohms. The critical frequency of the output circuit is,

Fc(output) = 1 / (6.28)(5E3)(20E-6)
= 1.6Hz

and is the circuit's dominant lower critical frequency (F1).

We can expect PSpice simulation and analysis of this circuit to display a bandwidth of approximately 2Hz to 45KHz. Gain should be the design estimate of 22.2. Next, we will enter the netlist in the circuit file.

Entering The Circuit Netlist

Observe circuit nodes and enter the ";notes" as you program PSpice to simulate the circuit. Include the values of the feedback divider, R1 and R2.

At Control Shell,
Select Files, Enter, Current File, and Enter.
Filename: LINBWAMP

The Netlist

```
*LINBWAMP - OP AMP BANDWIDTH
VCC  7 0 DC 15V          ;positive power supply
VEE  9 0 DC -15V         ;negative power supply
Vin  1 0 AC .1V SIN(0 0.1V 5KHz)    ;input AC sweep and stimulus
Rs  1 2 1KOhms           ;source resistance
Cin  2 5 10u             ;input coupling capacitor
R4  5 0 47K
R1  6 10                 ;feedback R
R2  6 11                 ;feedback divider R
Cby  11 0 50u            ;bypass capacitor
Cout  10 12 20u          ;output coupling capacitor
Rld  12 0 5K             ;load resistance
XAMP 5 6 7 9 10 UA741      ;"X" call for op amp UA741
.LIB EVAL.LIB            ;look for subcircuit in EVAL.LIB
.AC DEC 5 100m 200K        ;.AC sweep: .1Hz - 200KHz
.Tran .004ms .4ms 0 .002ms ;transient analysis, 2 hertz @ 5KHz
.Probe                   ;save data for Probe
.Options Nopage          ;minimize paging and banners in Output file
.End
```

Press Esc to save the circuit file.

Netlist Note

.Tran .004ms .4ms 0 .002ms

.Tran .004ms is the print_step interval for .print and .plot statements and is stored for Probe. In this statement, print_step is 1/100th of final_time, .4ms/100 = .004ms. Step_ceiling is calculated final_time/200 = .4ms/200 = .002ms. Default step_ceiling is final_time/50.

Run PSpice

Select Analysis, Enter, Run PSpice, and Enter.
Probe title screen should be displayed.

AC_sweep

Select AC_sweep, and Enter.
Select Add_trace, Enter, press F4.
Arrow to V(12), output load voltage, and Enter twice.
The sweep displays the expected bandwidth: 2Hz - 45KHz.

Note: Critical frequencies are at .707Vmax, at approximately 1.55V. The frequency and input stimulus are within specs, therefore, the display should be a representation of circuit bandwidth.

Maximum closed-loop voltage gain is near the estimated of 22, (2.19V / .1V = 21.9).

Select Remove_trace, and clear the screen for decibel gain measurements.
Select Add_trace, and type db(V(12)/V(1)), and Enter.
Select Y_axis, and Set_range: 0 30

Midband gain is near 26.8dB. (dB gain= (20)(log 21.9)= 26.81dB)
Select Auto_range, and Enter, then exit Y_axis menu screen.
Select Remove_trace, and clear the screen.

Select Add_trace, and again display the output, V(12).
Since the output coupling network has a higher critical frequency (1.6Hz), the limits of the input coupling network (.3Hz) and the bypass network (.7Hz) can be observed by probing V(10).
Select Add_trace, Enter, F4, V(10), and Enter twice.

The critical frequency for the combined input and bypass circuits is approximately the estimated .7Hz, limited by the bypass network.
Before leaving AC_sweep analysis, let's examine the input coupling circuit for the even lower critical frequency of .3Hz.
Select Remove_trace, and clear the screen.
Select Add_trace, F4, V(5), and Enter twice.

We can examine three characteristics of the input coupling network in the display. First, input coupling critical frequency is the lowest, as was expected, at .3Hz. Second, there is no limiting of higher frequencies in the RC coupling circuit. And finally, the level of the input signal at 100mV is not significantly attenuated by the input coupling circuit, indicating a stiff design.
Select Exit to leave AC_sweep analysis, and return to Probe title screen.

Transient_Analysis

Select Transient_analysis, and Enter.
Select Add_trace, Enter, F4, arrow to V(12), and Enter twice.
The waveform is undistorted, indicating that 5KHz is within circuit frequency limits and the input amplitude is not overdriving the circuit.
Gain is near the expected value of 22.2 and can be calculated with the output voltage at 2.2Vpeak. (Voltage gain= 2.2Vp / .1Vp = 22)

Overdriving the Input

As an experiment in overdriving the circuit, let's increase Vin, the input stimulus, and observe the results of transient analysis when we return to Probe.
Select Exit, Enter, Exit_program, and Enter.

Using the Stimulus Editor

At Control Shell,
Select StmEd, Enter, Edit, and Enter.

The program exits Control Shell and opens the Stimulus Editor. The screen displays the input stimulus.
Vpeak = 100mV, at 5KHz. (200us/hertz)

Select Modify_stimulus and Enter.
Select Transient_parameters and Enter.

Stimulus Editor adjustment options are VOFF (offset), VAMPL (amplitude), FREQ (frequency), TD (time delay), DF (damping factor), and PHASE.
Select 2)VAMPL, and Enter.
Backspace to erase, and type 1.35V, and Enter.
Select Exit, to leave the Transient_parameters menu.
The display indicates a new input stimulus of Vpeak= 1.35V. All other settings remain the same.

Select Exit, to leave the Modify_stimulus menu.
Select Exits, to leave StmEd.

Control Shell

Select Files, Enter, Edit, and Enter.
StmEd has written the modified stimulus into the circuit file. It reads,

```
Vin   1 0                        ;voltage source at Nodes 1 0
+AC .1V 0                        ;small-signal analysis input level
+SIN(0 1.350 5.000E3 0 0 0)      ;.Tran input: 1.35V, 5KHz
```

Stimulus values are:

SIN	Sine wave input stimulus
0	Voff, offset voltage
1.350	Input amplitude, Vpeak= 1.35V
5.000E3	Input frequency, 5KHz
0	TD, time delay
0	DF, damping factor
0	Phase

Press Esc, and save the edited netlist.

Run PSpice

Select Analysis, Enter, Run PSpice, and Enter.
At Probe title screen, select Transient_analysis, and Enter.
Select Add_trace, and display V(12).

The output waveform is displayed on the transient analysis graph. Severe clipping is occurring as the circuit is being driven into saturation and cutoff. The upper and lower limits of the power supplies are 15V and -15V, and are being exceeded.

Use StmEd to Reset Input Stimulus

Select Exit to leave Transient_analysis.
Select Exit_program to leave Probe.
Select StmEd at Control Shell, Enter, Edit, and Enter to access the Stimulus Editor.

In StmEd

Select Modify_stimulus, and Enter.
Select Transient_parameters, and Enter.
Select 2)VAMPL, and Enter.

Backspace and erase 1.350V, and type the original .1V, and Enter.
Select Exit, and Enter, to accept Transient_parameter changes. The modified stimulus, at 100mV, 2 hertz at 5KHz, is graphed.

Select Exit, and Enter, to leave the Modify_stimulus menu.
Select Exit, Enter, Exit_program, and Enter, to leave StmEd.

Review the circuit file with the Circuit Editor to assure that StmEd has made the necessary changes in the netlist.
Press Esc and save the file.

Run PSpice to check for original circuit performance.

Exit Probe.

Conclusion of Chapter 18

EOC Problems

(1) Using StmEd, adjust the stimulus amplitude to find the level where waveform clipping begins.

(2) Using StmEd, set transient analysis input amplitude to the minimum (sensitivity) value that will produce an undistorted output. Make gain calculations at each level of testing.

Chapter 19

SCHMITT TRIGGER CIRCUITS

Objectives: (1) To use PSpice and Probe to evaluate Schmitt trigger circuits
(2) To edit using options in PSpice Control Shell

I. Inverting Schmitt Trigger

A Schmitt trigger is a two-state circuit driven by an input voltage. It is triggered between states by the rising and falling edges of any input...sine wave, sawtooth, square wave or noise. In this exercise, the stimulus is applied to the inverting input of a 741 op amp. (Figure 1) Positive feedback to the non-inverting input enhances switching speed. A sine wave input will be used to exceed the trip levels of the circuit and produce a near square wave output.

The feedback network, R1 and R2, aids the input with a fraction of the output voltage that increases the input differential. When using the inverting input, the output is the opposite polarity and is fed back to the non-inverting input, driving the Schmitt trigger further toward saturation. From previous theory, the feedback fraction, B, is,

B = R2 / (R2 + R1)

In this exercise,

B = 15K / (15K + 150K) = .09

When the Schmitt trigger is saturated, a reference voltage is fed back to the non-inverting input and holds saturation. Saturation voltage is Vsat, and the reference voltage is,

Vref = BVsat   .09 (14.8V) = 1.33V

This reference voltage at the non-inverting input must be exceeded by the input voltage to effect a change of state in the Schmitt circuit. The reference voltage levels are the same when switching from either state, positive or negative. These input levels are called trip points.

Upper trip point,	Lower trip point,
UTP = BVsat	LTP = -BVsat
1.33V	-1.33 V

Hysteresis is a circuit quantity (and quality) calculated,

H = UTP - LTP = BVsat - (-BVsat)

Positive feedback creates the hysteresis and makes the circuit less sensitive to false triggering by noise or other voltage irregularities.

The saturation levels of the Schmitt trigger circuit will be measured by Probe, but first, the simulation and analysis by PSpice.

Enter the following circuit description and control statements with the Circuit Editor.

Filename: SCHMITT

The Netlist

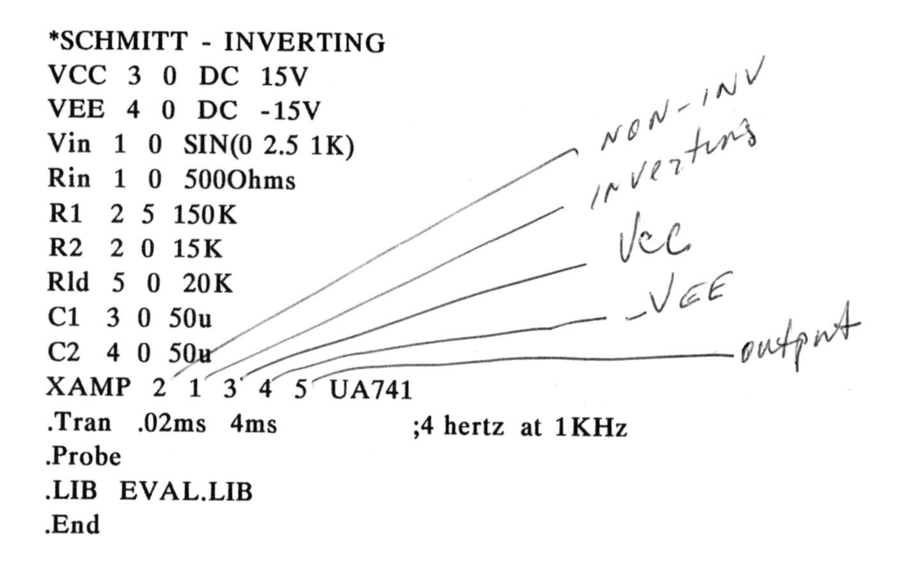

```
*SCHMITT - INVERTING
VCC  3  0  DC  15V
VEE  4  0  DC  -15V
Vin  1  0  SIN(0 2.5 1K)
Rin  1  0  500Ohms
R1   2  5  150K
R2   2  0  15K
Rld  5  0  20K
C1   3  0  50u
C2   4  0  50u
XAMP  2  1  3  4  5  UA741
.Tran  .02ms  4ms            ;4 hertz at 1KHz
.Probe
.LIB  EVAL.LIB
.End
```


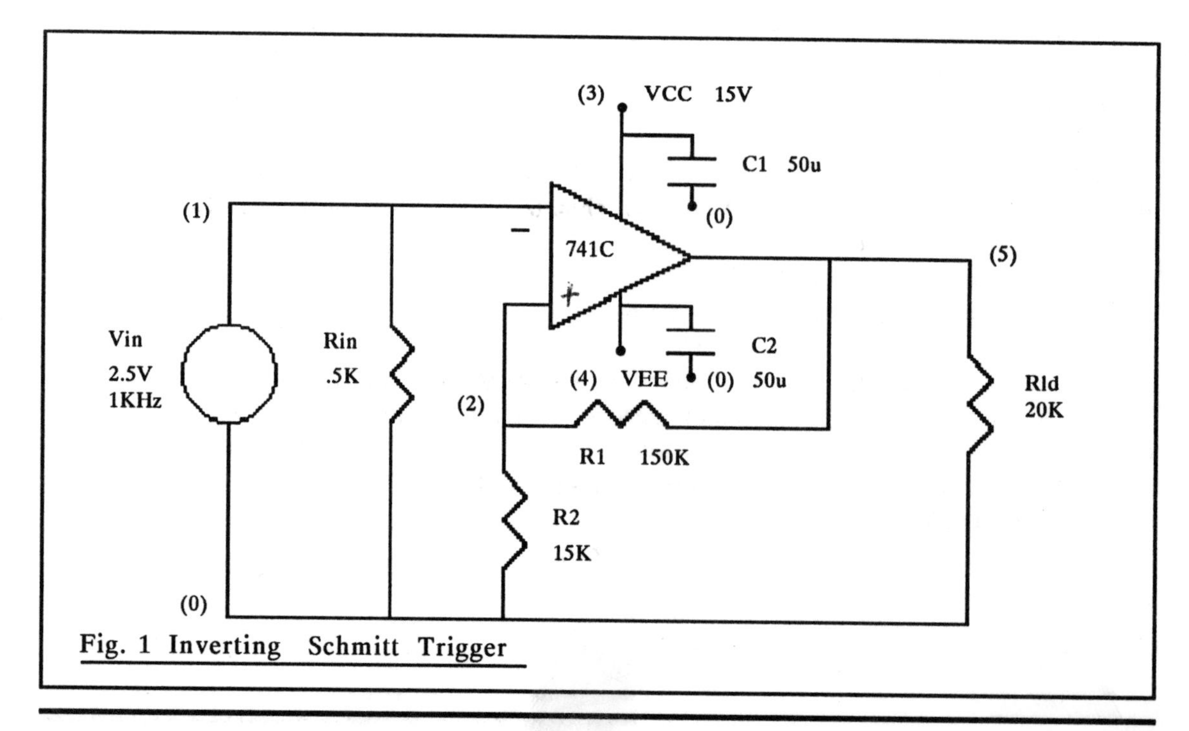

Fig. 1 Inverting Schmitt Trigger

Browse the Control Shell

Let's take a look around the Control Shell before running PSpice.

Press Esc and save the circuit file.

Select Circuit, Enter, Devices, and Enter.

A list of circuit devices is presented in the Circuit pop-down menu. Devices can be changed by using the Arrow keys and pressing Enter to select. Component models may be specified for various analyses. The XAMP 741 op amp cannot be modified at this screen because it is a subcircuit written in the library. Those parameters are not adjustable in the Control Shell.

Edit C1 and C2

Edit C1 and C2, the 50ufd capacitors, to 40ufd.

Arrow to C2, and Enter.

At C2 screen, arrow to the capacitance line, and Enter.

At New value?, type 40u , and Enter.

Press Esc to exit and save C2 changes.

Repeat for C1, changing its value to 40u.

Press Esc to exit.

Next, take a look at C1 and C2 changes that were edited into the circuit file. Select Files, Enter, Edit, and Enter.

Press Esc to exit the Circuit Editor.

StmEd

Select StmEd, Enter, Edit, and Enter.

Note that PSpice needs your OK to exit Control Shell and accept changes made in the circuit file. Changes are saved on disk.

Accept "S" and Enter.

StmEd's initial menu screen is displayed. Four hertz of the 1KHz, 2.5Vpeak, input stimulus are graphed for editing. All changes made with StmEd are also written in the circuit file.

Select Modify_stimulus, and Enter.

Select Transient_parameters, and Enter.

Six parameters are presented for editing: Offset voltage, amplitude, frequency, time delay, damping factor, and phase. A table displays the present values.

Modify stimulus frequency to 500Hz and amplitude to 2.25V.

Select 3)FREQ, and Enter.

Backspace to erase, and type .5E3, then Enter.

Select 2)AMPL, and set amplitude to 2.25V.

Select Exit, and Enter.

The new stimulus is displayed. Frequency has been edited to 500Hz and the amplitude is 2.25V. The graph is set for transient_analysis final_time of 4ms and covers 2 hertz at 500Hz.

Select Exit, and Enter, to leave Modify_stimulus screen.

Select Exit, and Enter, to leave StmEd initial menu screen.

Select Exit_program, and Enter, to leave StmEd.

Using the Circuit Editor, take a look at the new frequency and amplitude that has been edited into the netlist by StmEd.

Press Esc to exit.

Browse Analysis Pop-Down Menu

Select Analysis, and Enter.

Select Transient..., and Enter.

The pop-down menu screen provides easy editing of the transient analysis input statement. Transient analysis is in the Enable position. A Disable "N" would suspend this analysis on the next run. Print_step is set to 20.000us. Final time is at 4.000ms (2 hertz at 500Hz). Output No-print Delay is set to 0. Step Ceiling is 0, and accepts the default value of final_time/50. Detailed bias pt is "N," and Use Initial Condition (UIC) is "N." No Fourier Analysis has been specified, therefore, "N" is indicated.

Edit final_time to 6ms and No-print delay (results_delay) to .5ms.

Press Enter for "Y" to accept Enable Transient mode.

Arrow to Final Time.

Type 6.0ms , and Enter. The new final time of 6ms specifies that transient analysis of the circuit last for 3 hertz at 500Hz.

At No-print delay, type .5ms , and Enter. The output will be delayed for the first one-quarter hertz of the 3 hertz transient analysis. (No-print delay is results_delay, the third entry in .Tran statements)

Continue to press Enter, and accept all other settings.

Accept "Y" for Done? , and Enter.

Press Esc to return to Control Shell.

Other Analysis Options

Select Analysis, and Enter.

Select each of the options listed, except Run PSpice, and browse. When finished, press Esc to leave these edit screens.

AC & Noise..., DC Sweep..., Parametric..., and Monte Carlo... have not been enabled or specified for the present circuit. The temperature setting will remain at default 27 degrees C (no temperature specified), and the Change Options screen provides a list of options available.

Press Esc to return to Control Shell.

Select Files, and Edit, and take a look at the .Tran statement for changes in final_time and results_delay that have been edited into the circuit file.

Press Esc to exit.

Next, Run PSpice simulation and analysis of the Schmitt trigger circuit.

Run PSpice

Select Analysis, and Enter
Select Run PSpice, and Enter

The initial Probe menu screen is displayed.
Select Add_trace, and Enter.
Press F4 for a list of variables to view.
Arrow to V(5) and Enter twice.
V(5) is the Schmitt trigger output, switching between the two saturation levels.

Note the quarter hertz delay in the waveform. Next, notice the saturation level of the output. A more accurate measurement can be made if the Y_axis range is reset.

Select Y_axis, and Enter.
Select Set_range, and Enter.
Saturation swings can be no more than the supply voltages, therefore, at Enter a range: type -15V 15V , and Enter.

Maximum levels are -14.6V to 14.6V.

Select Auto_range, and Enter.
Select Exit, and Enter to return to the initial Probe menu screen.
Select Remove_trace, and clear the screen.

Select Add_trace, Enter, and press F4.
Arrow to V(1), the inverting input, and Enter twice.

The input voltage is 2.25Vpeak, 2 3/4 hertz of the 500Hz stimulus, delayed 1/4 hertz.

Display Input and Output

Select Add_trace, and Enter.
Press F4, arrow to V(5), and Enter twice.

The screen displays both the input and Schmitt trigger output. But where exactly are the "trip points" of the input? Both upper and lower trip levels are slightly above the reference voltage, which is,

Vref = BVsat
Vref = .09(14.6) = 1.314 Volts

Calculate circuit hysteresis by inserting values for UTP and LTP,

H = UTP - LTP
H = BVsat - (-BVsat)
H = 2.628Volts

How can we prove the trip points? If we display the Schmitt trigger output on the Y_axis with the input stimulus on the X_axis, circuit hysteresis will be graphed.

Graph Hysteresis

Select Remove_trace, and clear the screen.
Select X_axis, and Enter.
Select X_variable, and Enter.
Press F4 for a list of variables to use on the X_axis.
Arrow to the input, V(1), and Enter twice.
The X_axis range is set to -3.0V to 3.0V, to include the 2.25V stimulus amplitude.
Select Exit, and return to the initial Probe menu screen.
Select Add_trace, F4, arrow to V(5), and Enter twice.
Output versus input is a graph of the circuit hysteresis loop. Trip levels are readable at slightly above the reference level of 1.314V on the X_axis. Trip points are at the upper right and lower left corners of the trace. Hysteresis is the voltage difference between these two points, and must be just above the estimated 2.628V.
Exit Probe and return to the Control Shell.

Inverting Schmitt Problems

(1) The following measurements will verify circuit trip points. Edit the stimulus amplitude to 1.2Volts (just below the 1.314V trip level). Run PSpice and check for triggering. Display both V(1) and V(5). V(5) should be holding saturation. Repeat with an input of 1.4V and run the same test. Use the StmEd program.

Note: The test should show a fixed saturation level, no triggering with 1.2V input, and stable operation with 1.4V input.

(2) Change R1 and R2 values, calculate new trip point values, and Run PSpice to verify your estimates.

II. Non-Inverting Schmitt Trigger

The Schmitt Trigger circuit analyzed in the first section of the chapter was triggered with a stimulus applied to the inverting input of the op amp. Trip level was 1.314V. In this section, the non-inverting input will be used with the 741, and the trip level is near 0V at the non-inverting pin.

Enter the following netlist. Filename: SCHMITT2

```
*SCHMITT - NON-INVERTING
VCC  3 0 DC 15V
VEE  4 0 DC -15V
Vin  5 0 SIN(0 .15 1K)
R2   5 1 1.5K
R1   1 7 150K
Rld  7 0 22K
XAMP 1 0 3 4 7 UA741
.Tran .02ms 4ms 0 .04ms
.Probe
.LIB EVAL.LIB
.End
```

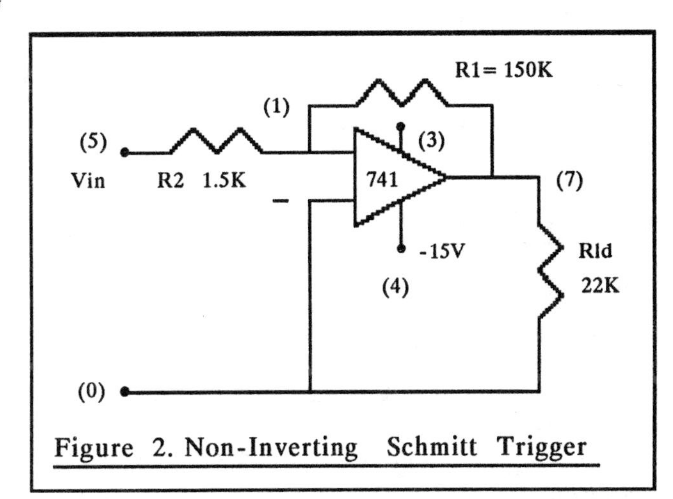

Figure 2. Non-Inverting Schmitt Trigger

Press Esc to save the file.

Circuit Estimates

Trip Points

Vsat = 14.6V

UTP = (R2/R1) (Vsat) = .146V
LTP = -(R2/R1) (Vsat) = -.146V

Hystereris

H = UTP - LTP
= .292V

Run PSpice

At Probe

Select Add_trace, and measure the output waveform at V(7).
Select Remove_trace, and clear the screen.

Select Add_trace, and examine the input stimulus, V(5).
Select Remove_trace, and clear the screen.

Graph Hysteresis

Set V(5), the input voltage, as the X_axis variable.
Select X_axis, and Enter.
Select X_variable, and Enter.
Press F4, arrow to V(5), and Enter twice.
Select Exit, and Enter.

Select Add_trace, arrow to V(7), the output voltage, and Enter twice.
The circuit hysteresis loop is displayed.
Select X_axis, Set_range, and set range to: -200mV 200mV.
High to low switching is at the upper left of the hysteresis loop and low to high switching is at the lower right.
Trip levels are near .146V and -.146V.
Select Auto_range, and Enter.
Select Exit, and Enter.
Select Remove_trace, and clear the screen in preparation for graphing a hysteresis of output voltage versus input at the non-inverting pin.
Select X_axis, Enter, select X_variable, and Enter.
Press F4, select V(1), and Enter twice.
Select Exit, and Enter.
Select Add_trace, arrow to V(7), and Enter twice.

Note the zero center voltage at the non-inverting pin input.
Select Exit, and Enter, to leave Probe.

Set the input stimulus amplitude to .14V and repeat measurements to illustrate that this amplitude is below trip points, .146V and -.146V, and will not trigger the circuit.
Reset the input amplitude to .15V and test for triggering.

Non-Inverting Schmitt Problem

(1) Experiment with the non-inverting Schmitt trigger circuit by changing component values and repeating measurements.

Chapter 20

MORE PSPICE CIRCUITS

I. Op Amp Relaxation Oscillator

The 741 op amp can be used as a relaxation oscillator and produce a square wave output. The circuit utilizes a feedback divider similiar to the one used in the Schmitt trigger. (Figure 1) R1 and R2 provide the feedback fraction, B. The time constant of the RC network and the following equation solve for the time period of one cycle of the square wave output.

T(time)= 2RC(ln (1+B/1-B))

(The natural logarithm, to the base e, is used in the equation. If your calculator has only common logarithm, multiply by 2.3026 for natural logarithm)

B= R2 / (R2 + R1) = 27K / (27K + 3K) = .9

RC= (1.7K)(.1ufd) = .17ms

T= (2)(.17ms)(ln(1+.9/1-.9)) = 1ms

F= 1/T = 1KHz

Review the circuit in Figure 1, and enter the netlist.
Filename: RELAXATI

```
*OP AMP RELAXATION  OSC
VCC  3  0  DC  15V
VCC  8  0  DC  -15V
C1     5  0   .1u
R3     5  7  1.7K
R1     7  4  3K
R2     4  0  27K
XAMP  4  5  3  8  7  UA741
.Tran  .02ms  2ms  0  .04ms  UIC
.IC  V(5)=0V  V(7)=0V
.LIB  EVAL.LIB
.Probe
.End
```

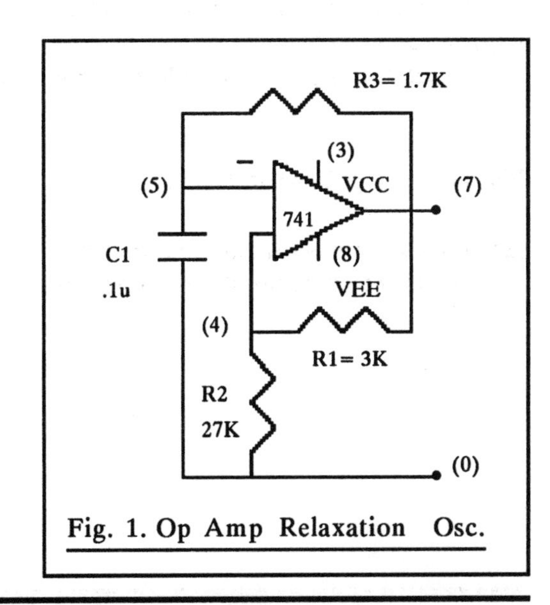

Fig. 1. Op Amp Relaxation Osc.

Netlist Note

.IC, Initial Bias-Point Condition, enables the PSpice user to specify a particular circuit node voltage (or device current) when transient analysis and/or small-signal analysis begins. .IC can also be used in capacitor statements, i.e., C4 6 3 .01u IC=initial_voltage, and inductor statements. .Nodeset provides initial voltage estimates and is available in all PSpice analyses. .IC overrides .Nodeset for transient analysis only when both are used at the same point in the circuit.

.Nodeset is also used to define a start point for the analysis of circuits with more than one stable condition, i.e., flip-flops.

Use Initial Condition, UIC, at the end of .Tran, suspends bias point calculations. This option must be used when the Initial Bias-Point Condition, .IC, is specified. .IC V(5)=0V V(7)=0V sets these two node voltages as the initial condition for transient bias points.

Run PSpice

Probe

Examine V(7), for the 1KHz square wave output.

Examine the charge and discharge waveform on C1 at V(5). UTP and LTP are at the extremes of this charge/discharge curve.

Displaying V(7) and V(5) together illustrates triggering the input at V(5).

Section Problems

(1) Redesign the circuit for an output of 500Hz.

(2) Set component values: R1= 4.7K, R2= 33K, C1= 10ufd, and R3= 2K.
Set transient analysis times: .Tran .02s 2s 1.4s .02s UIC
Results_delay is set at 1.4s and delays transient analysis reporting by this time. Note this delay when viewing V(7). Calculate output frequency and Run PSpice to verify estimates.

(3) Reduce the resistor values in Problem 2 by one half. Recalculate frequency and Run PSpice to check predictions. (Frequency should double)

II. JFET Amplifier

We examined the 2N3819 JFET in Chapter 9. PSpice was used to draw a transconductance curve and IDSS was measured at approximately 11.8mA. VGS(off)= -3V. The following JFET amplifier uses these quantities and, with the aid of the same transconductance curve, the source resistor has been set to 300 Ohms to bias the transistor for optimum operation. With Rs= 300 Ohms, ID= ~4mA, and VGS= approximately -1.2V.

Enter the following netlist after reviewing the circuit in Figure 2.
Filename: JFETAMP

```
*JFET AMPLIFIER
VDD 1 0 DC 12V
Vin 5 0 AC 1 SIN(0 .2V 1KHz)
Cin 5 4 .022u
Rg 4 0 1MEG
Rd 1 2 1.5K
Rs 3 0 300Ohms
*Cby 3 0 27u
J1 2 4 3 J2N3819
.Model J2N3819 NJF(Vto=-3 Is=33.57f Beta=1.304m)
.Tran .02m 2m 0 .01m
.AC DEC 5 1m 100k
.LIB EVAL.LIB
.Probe
.End
```

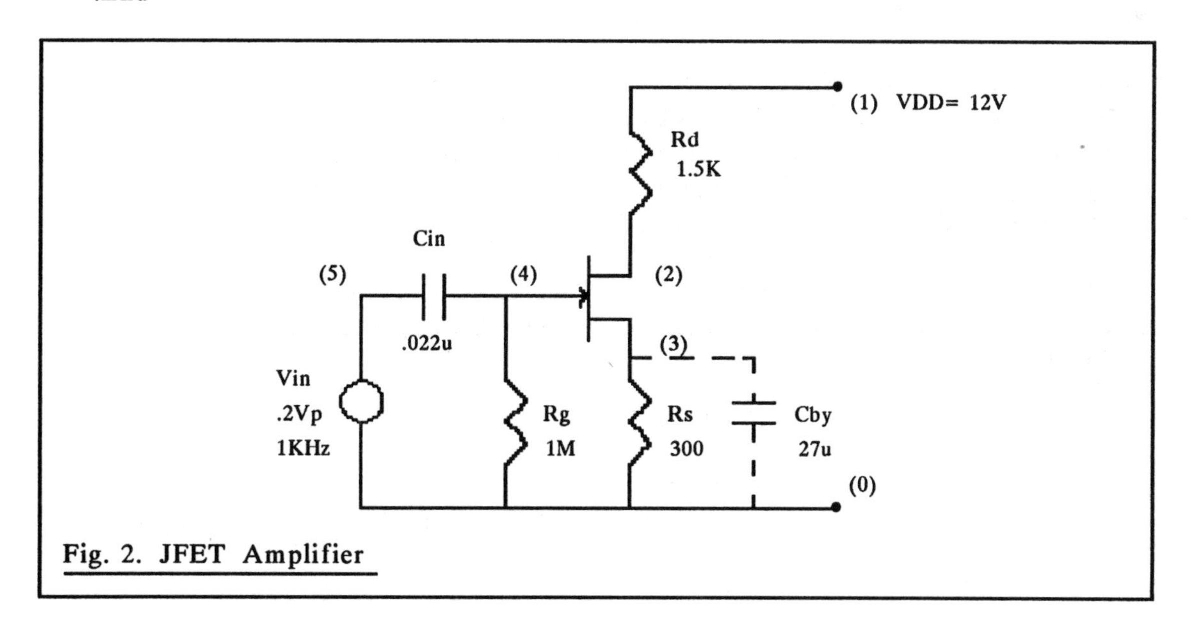

Fig. 2. JFET Amplifier

Press Exc and save the circuit file.

Netlist Notes

Drain resistor, Rd, has been set to drop one-half VDD. Gain is estimated at 3 with Cby disconnected, and near 7 with the capacitor in the source circuit.

Using your theory textbook as a guide, complete circuit performance estimates before running PSpice.

Run PSpice

Run PSpice simulation and analysis and compare Output file data and Probe analysis with circuit calculations.

Check circuit saturation by increasing Vin and Run PSpice analysis. Measure gain at each setting of Vin.

Section Problems

(1) Select JFET amplifier circuits from theory text and, after completing calculations, Run PSpice to check the operation of these circuits.

(2) Design a JFET amplifier to your specifications. Run PSpice to verify calculations.

Note: Two N-channel JFETs are included in the Evaluation Library; J2N3819 and J2N4393.

III. Colpitts Oscillator

A Colpitts oscillator is presented in the netlist below and in the circuit on the next page. Oscillator frequency is determined by,

$$F = 1 / (6.28) (\sqrt{LC})$$

where,

$$C = (C1)(C2)/(C1+C2)$$

Calculated frequency is 71.2KHz.

The feedback fraction, feedback voltage/output voltage, is determined by the selection of C1 and C2, the capacitors used when setting resonant frequency. The Vfeedback/Vout ratio is the same as the ratio of these two capacitors used as a voltage divider,

$$\text{Feedback fraction} = C1 \;/\; (C1 + C2)$$

$$= .0022u \;/\; (.0022u + .022u)$$

$$= .091$$

Enter the netlist at the Circuit Editor.
Filename: COLPITTS

```
*COLPITTS OSC
VCC  4  0  DC 12V
R1  4  2  40K
R2  2  0  6.8K
Cby  2  0  .01u
L1  4  6  2.5mh
C3  4  0  .1u
C1  6  3  .0022u
C2  3  0  .022u
RE  3  0  14.7K
Q1  6  2  3  Q2N3904
.Model  Q2N3904  NPN(bf=100 cjc=10p rb=20)
.Tran  .28us  56us  0  .56us  UIC    ;4 hertz at 71.2KHz
.IC  V(3)=0V
.LIB  EVAL.LIB
.Probe
.End
```

Press Esc and save the circuit file.

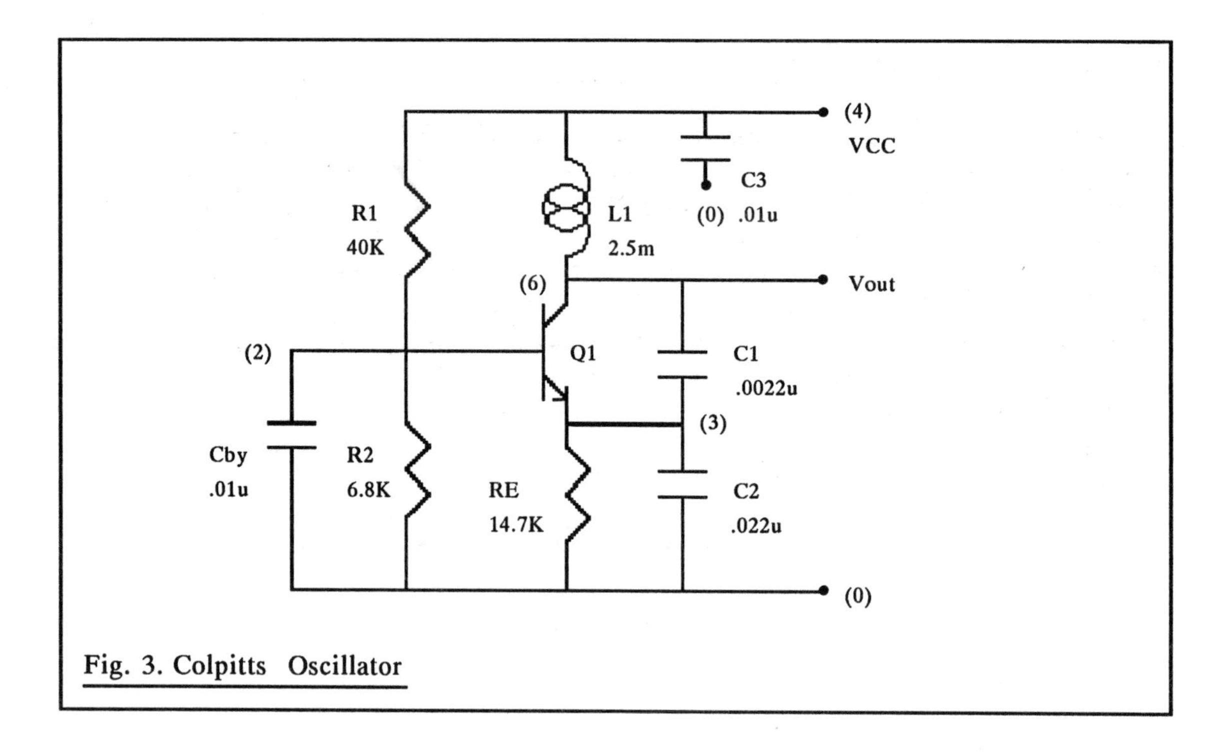

Fig. 3. Colpitts Oscillator

Netlist Notes

Capacitor C3 grounds the oscillator signal and prevents coupling to other circuits using the same voltage source. Cby bypasses the base to ground and improves gain. The output voltage is fed back to the emitter through the capacitor divider network.

The UIC entry at the end of the .Tran statement suspends bias point calculations. Place an asterisk before the .Tran statement and the .IC line and Run PSpice to measure transistor bias voltages.

Run PSpice

Remove asterisks and Run PSpice analysis to test circuit operation.

Run PSpice

Test Measurements:

(1) Measure output at V(6) for amplitude and frequency expected.
(2) Measure feedback voltage at V(3) and calculate feedback fraction, Vfeedback/Vout, to check calculations.

Section Problem

(1) Redesign the oscillator circuit for frequencies of 50KHz and 100KHz. Keep the same feedback fraction, .09, in each new design.

IV. Complementary Symmetry Amplifier

The complementary symmetry amplifier is used in circuits requiring more power and higher efficiency than is available in single transistor Class A amplifiers. One disadvantage of this amplifier is "crossover distortion." This distortion of the output waveform occurs when the input signal passes through zero volts. Diodes are placed in the circuit to bias the output transistors into operation to reduce crossover.

Enter the netlist below and consult your theory text for circuit operation details. The circuit diagram is Figure 4 on the next page.

Filename: COMSYAMP

```
*COMPLEMENTARY SYMMETRY AMPLIFIER
VCC  4  0  DC 14V
Vin  1  0  SIN(0 1.5V 1K)
Cin  1  2  8u
D1   5  2  D1N4001
D2   2  7  D1N4001
R1   4  5  12K
R2   7  0  12K
Cout  3  8  540u
Rld  8  0  100Ohms
Q1  4  5  3  Q2N3904
Q2  0  7  3  Q2N3906
.Model  D1N4001  D(Is=10.0E-15 Rs=.1 Ikf=0 Cjo=1p N=1
+                 Eg=1.11  Xti=3 Vj=.75 Fc=.5 Nr=2 Bv=100
+                 Ibv=100.0E-6  Tt=5n  Isr=100.0E-12)
.Model  Q2N3904  NPN
.Model  Q2N3906  PNP
.Tran  .04ms  2ms  0  .02ms
.LIB  EVAL.LIB
.Probe
.End
```

Press Esc and save the circuit file.

Netlist Notes

The diode model is not included in the Evaluation Library and must be written into the netlist.

Q1 and Q2 are biased on by the voltages across the 1N4001 diodes but this biasing does not completely remove the crossover problem. These power diodes do not match the transistor emitter-base diodes. The output at V(8) will display the distortion.

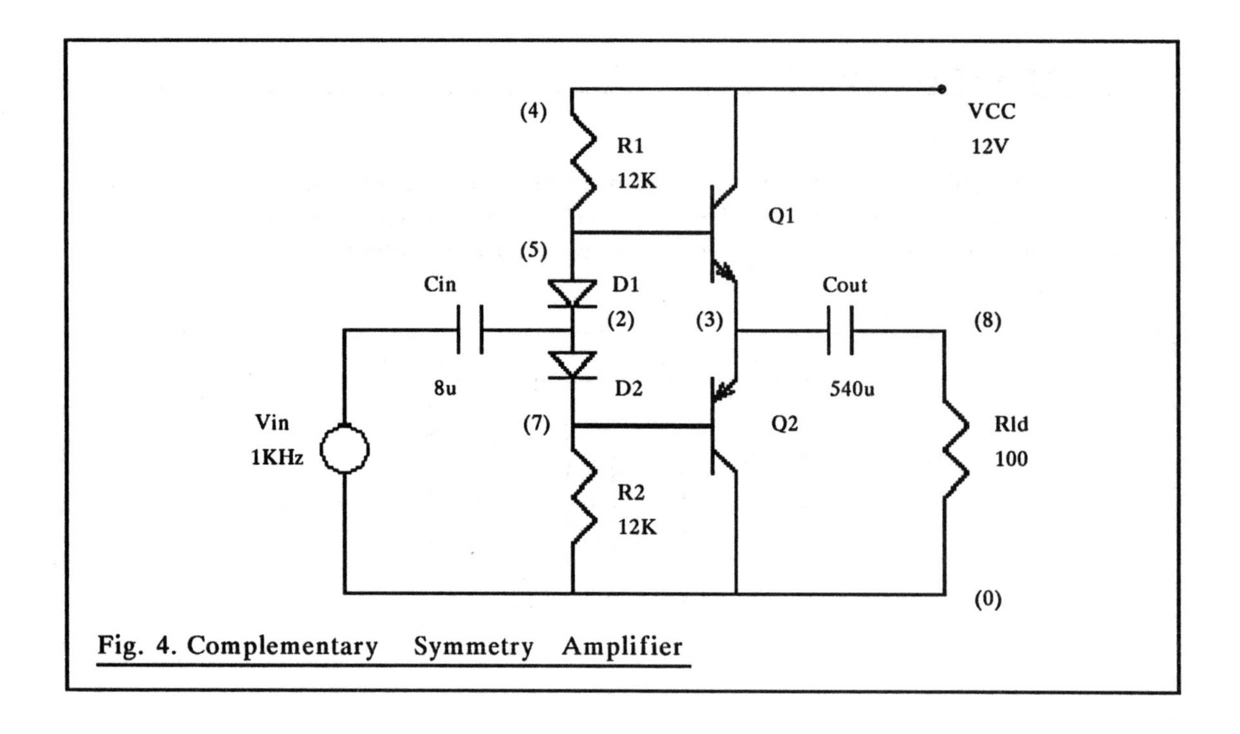

Fig. 4. Complementary Symmetry Amplifier

Run PSpice

Test Measurements:

(1) Examine V(8) for crossover distortion.
(2) Decrease the stimulus to .5Vpeak and repeat the V(8) measurement. Note that distortion is more pronounced with the lower input.
(3) In an attempt to match the transistor emitter-base diodes, replace the two 1N4001 diodes with two Q2N3904 transistors connected as diodes. (Q3 5 5 2 Q2N3904 and Q4 2 2 7 Q2N3904) Repeat the measurement at V(8).
    Increase the stimulus to 2.5Vpeak and examine V(8) for distortion with increased input.
(4) Increase the stimulus to a level that saturates the amplifier. Examine V(8), V(5), and V(7) simultaneously and compare distortion.

Section Problem

(1) Simulate and analyze complementary symmetry amplifiers that you have studied in theory text.

V. Active Low-Pass Filter

In theory studies, op amp circuits have been introduced as inexpensive, space conserving, active filters. Review the "two-pole" low-pass filter circuit in Figure 5, and enter the netlist in preparation for examining output design requirements.

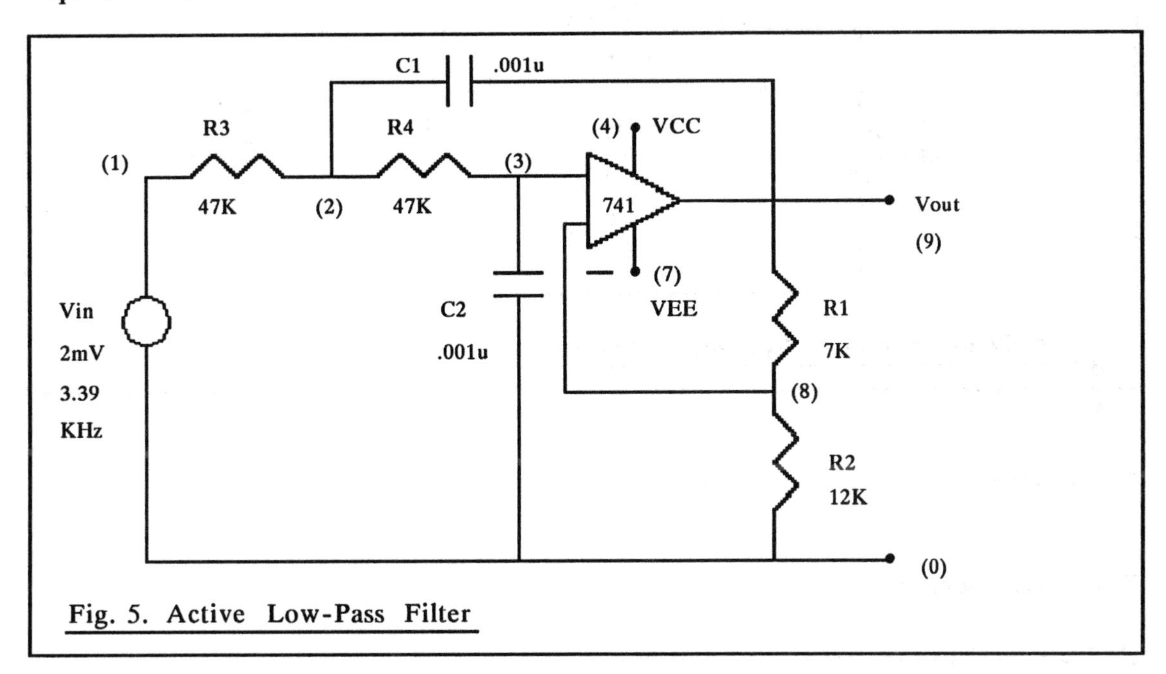

Fig. 5. Active Low-Pass Filter

Filename: BUTTERWORTH

```
*ACTIVE LOW-PASS FILTER
VCC  4  0  DC 15V
VEE  7  0  DC -15V
Vin  1  0  AC .002  SIN(0 2mV 3.39KHz)
R1  9 8  7K
R2  8  0  12K
C1  2  9  .001u
C2  3  0  .001u
R3  1  2  47K
R4  2  3  47K
XAMP  3  8  4  7  9  UA741
.Tran  .022ms  1.12ms  0  .012ms
.AC  DEC  5  1  100K
.LIB  EVAL.LIB
.Probe
.End
```

Circuit Calculations

Midband closed-loop voltage gain is,

Acl = (R1 / R2) + 1
= (7K / 12K) + 1
= 1.58

Output voltage at midband frequencies,

Vout = (Acl)(Vin)
= (1.58)(2mV)
= 3.16mVp

Low-pass filter cutoff frequency, when Acl is 1.586,

F = 1 / (6.28)(RC)
= 1 / (6.28)(47KOhms)(.001ufd)
= 3.39KHz

The output voltage is 3dB down at the cutoff frequency.

Vout(cutoff) = (.707)(3.16mV)
= 2.23mVp

A closed-loop voltage gain of 1.58 is near the required Butterworth value of 1.586. Midband gain should be as constant (flat) as possible.

Run PSpice

Select AC_sweep at Probe's title screen and press Enter.
Select Add_trace, and measure the response of the low-pass filter at V((9). Note the 3dB level at the cutoff frequency, 3.39KHz.
Select Exit, and return to Probe's title screen.
Select Transient_analysis, and Enter.
Select Add_trace, and measure the sine wave output at V(9).
The output should be at the estimated 2.23mVp at the cutoff frequency.

Section Problems

(1) Output voltage decreases at a rate of approximately 40dB/decade above the cutoff frequency. Set input to 5KHz, 6KHz, and 10KHz and measure gain to confirm this estimate.
(2) Redesign the filter for a different cutoff frequency. Use different resistor values but maintain Butterworth requirement of 1.586 gain.

SETUPDEV.EXE PROGRAM

If it becomes necessary to change your system...different monitor or printer, for example, the SETUPDEV program in PSEVAL52 is available. The program is menu driven and allows quick access and egress.

At the PSEVAL52 directory, type,

SETUPDEV, and Enter.

The screen should display the initial SETUPDEV menu.

```
SETUPDEV  - PSPICE Device file creation program.  Version 5.2

Device file : pspice.dev
_____________________________

Display   =
Hard-Copy  =

0. Exit
1. Change Display
2. Add Hard-Copy Device
3. Delete Hard-Copy Device
4. Save Device File
Selection>
```

Select the number that designates the device you wish to change and press Enter.

At the device options window, choose the number that designates your monitor or printer. Press Enter to make the change.

Select (3) to Delete Hard-Copy Devices.

Select (4) to Save Device File.

Select (0) to Exit the SETUPDEV program.

Conclusion of SETUPDEV program.

COMPLIMENTARY Evaluation Version Software

Instructors may obtain a complimentary copy of PSpice Evaluation Version 5.2 or 5.3 software from C>PAT Publishers at the address listed below. Please submit your request on school letterhead to:

Computer Programming And Teaching Publishers
P.O. Box 351
Henderson, NC 27536

BIBLIOGRAPHY

MicroSim PSpice(C) and Design System 2(C):

Evaluation - Version 4.03 program, 1/22/90,

Evaluation - Version 4.04 program, 7/10/90

Evaluation - Version 4.04b program, 9/26/90

Evaluation - Version 4.05 program, 2/21/91

Evaluation - Version 5.0 program, 1991

Evaluation - Version 5.2 program, 1992

Evaluation - Version 5.3 program, 1993

MicroSim Corporation, 20 Fairbanks, Irvine,
California 92718 USA

SPICE, SPICE2G, SPICE2G.6, SPICE3F
Simulation Program with Integrated Circuit Emphasis,
Industrial Liaison Program, Department of Electrical Engineering and
Computer Sciences/Electronics Research Laboratory, 231 Cory Hall, University
of California at Berkeley, Berkeley, CA. 94720

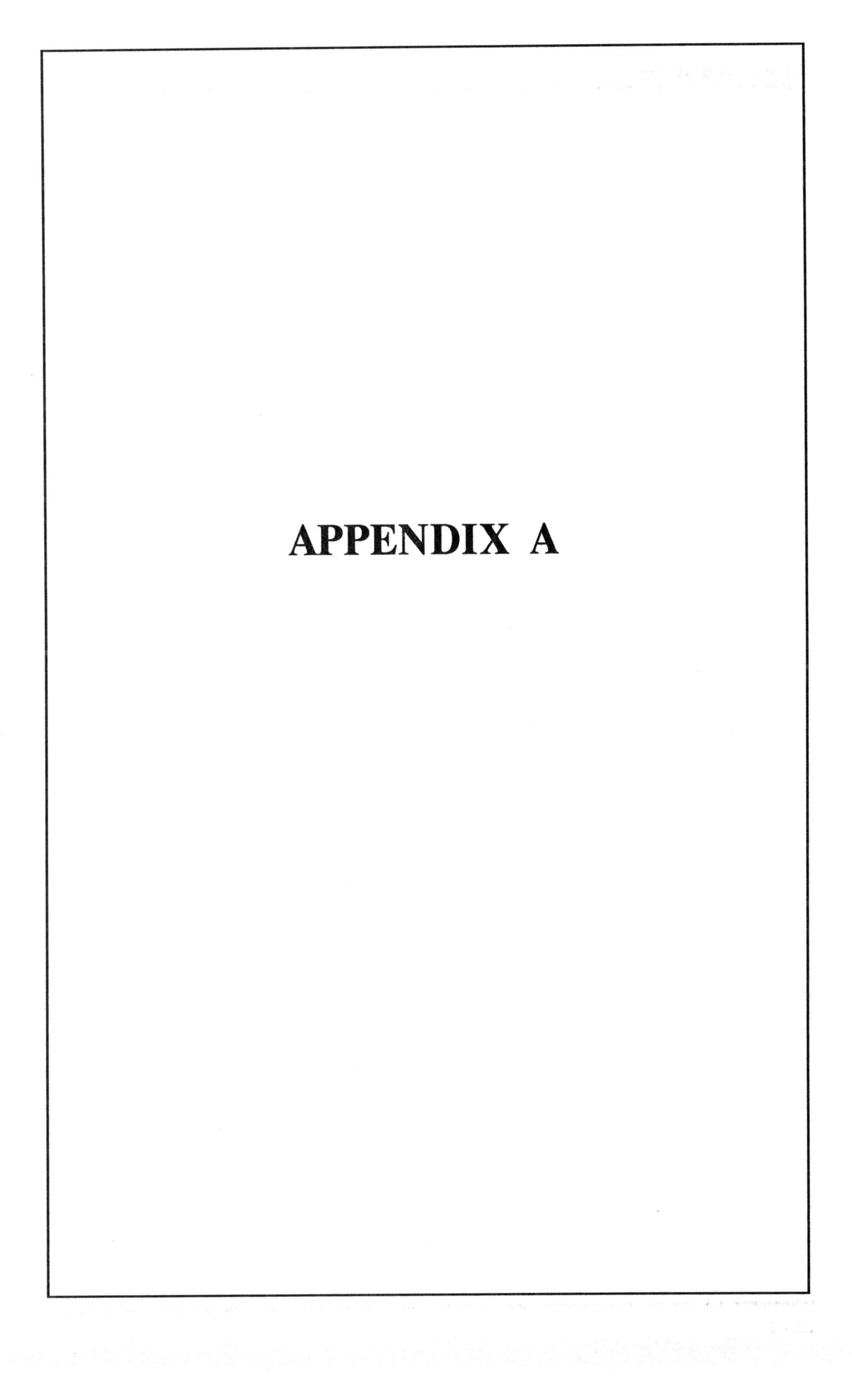

APPENDIX A

APPENDIX A

APPENDIX A

SPICE STATEMENTS

An introductory list of notations and control statements used in Spice programming is presented in Appendix A.

First Line

The first line in the netlist is the circuit title line, with or without an asterisk.

Second Line

The second line begins the circuit netlist statements.

* An asterisk (*) at the beginning of a line indicates a comment.
*For example, this line would be ignored by PSpice.

; The semicolon indicates an in-line comment ;i.e. PSpice ignores this comment

.AC Small-signal analysis

.AC {sweep type: DEC LIN OCT} points start_frequency stop_frequency
The .AC sweep statement provides a small-signal frequency response analysis of the circuit. This statement will sweep the circuit input, Vin for example, with the frequencies specified. The Probe sweep analysis program will display the outputs requested.

Input levels for .AC analysis are defined in the source line, for example, Vin 1 0 AC 1. AC 1 identifies the input of 1Volt or unity input to be used in small-signal analysis. When .AC sweep gain measurements are in reference to unity input, the output is read as gain, not the sometimes unrealitic voltage levels. The high input level of 1Volt in some circuits is a reference level only and does not disturb gain measurements because PSpice has linearized the circuit and saved the linear equivalent of the circuit for .AC sweep analysis. Later in the netlist, the .AC sweep small-signal analysis control statement sets sweep frequency limits and the data sample number.

PSpice will perform three types of frequency definitions in the .AC sweep statement. They are best seen with examples:

.AC DEC 100 10K 1MEG. A decade sweep is requested with 100 points per decade. Decade is a 10X increase in frequency. The range: 10KHz - 1MEGHz

.AC LIN 200 10K 1MEG, defines a linear sweep with 200 sweep data points, from 10KHz to 1MEGHz.

.AC OCT 10 10K 1MEG, request a logarithmic sweep with the sweep points per octave as the first number in the statement, 10. Examples of octave increases in this statement are 10KHz, 20KHz, 40KHz, 80KHz, etc. The sweep range is 10KHz to 1MEGHz. (See Chapter 13 and note on page 16-16)

B GaAsFET, B(Name) (Drain) (Gate) (Source) (Model)
Gallium-arsenide field effect transistor.

C Capacitor, C(Name) (Node1) (Node2) (Size)
C2 1 0 10u

D Diode, D(Name) (Node1) (Node2) (Model)
D6 1 0 D1N4001

.DC

Provides a DC sweep of voltage, current, resistor models, or temperature.
For examples, let's look at sweeping source voltage or current. Typically, .DC sweeps start at one value of the source and calculates all circuit bias points, then moves to the next requested value and recalculates bias points again. The sweep used in drawing collector curves in Chapter 9 is,

.DC VCE 0 18 .1

Sources to be swept with the .DC sweep statement must first be entered into the circuit file. The source voltage, VCE, is listed in the netlist as a circuit description statement, for example, VCE 2 0. The line defines VCE connected at Nodes 2 and 0. This source will be swept in .1V increments, from 0V to 18V. Listed terms are: .DC (type of sweep), VCE (sweep variable), 0 (start voltage value), 18 (stop voltage value), and .1 (increment voltage value).
Types of sweep are linear (LIN), decade (DEC), octave (OCT), and sweep a list (LIST) of values that follow in the statement. Default is the linear sweep.
As the source is swept, measurements of voltage and current throughout the circuit can be printed and plotted in the .Output file with the .Print and .Plot statements. For examples,
.Print DC V(any node or two terminal device)
.Print DC I(any two terminal device)
.Plot DC V(specified node or two terminal device)
The print and plot outputs will include all the calculated sweep step values if the range is not limited.

More Examples of .DC Sweep

.DC Vpwr 0 12 1, sweeps the input voltage from 0 to 12 Volts in 1 Volt steps. Three type sweeps can be specified, Linear (LIN), Decade (DEC), and

Octave (OCT), and each will give a corresponding plot. The default sweep is linear.

.DC VCE 0 18 .1 IB 20E-6 100E-6 20E-6

PSpice DC analysis begins at a base current of 20uA, and steps 20uA to 100uA. Steps are 20uA, 40uA, 60uA, 80uA, and 100uA. Collector-emitter voltage will step from 0V to 18V in .1Volt steps for each value of base current.

The progress of the DC analysis is displayed during the test. The first "Start" line is base current. The second "Start" line is collector-emitter voltage steps.

.DC VDS 0 16 .4 VGS 0 -3 .5

This DC sweep statement will program PSpice to run a sweep of JFET drain voltage, VDS, for each value of gate voltage. Gate voltage will step from 0Volts to -3Volts in .5Volt steps. Drain current will be cutoff at the Vto level of -3Volts, the pinchoff voltage.

.DC VGS -0 -3.5 .1

Gate input voltage will be stepped in a JFET circuit from 0Volts to -3.5Volts in .1Volt increments. Drain voltage is constant.

Note: See page 7-10 for .DC sweep and resistor models.

.End

A necessary statement to signal the end of the circuit file. PSpice simulation and analysis begins only if the circuit file has this statement. Another circuit file can be entered after the .End statement.

.ENDS

End of subcircuit definition, started by .SUBCKT

.FOUR

Fourier analysis, .FOUR, accompanies a transient analysis, .TRAN, statement. A harmonic decomposition on a stated fundamental frequency, including the DC, fundamental, 2nd through 9th harmonic component, and harmonic distortion is analyzed. For example, in the following statement,

.FOUR 500 V(6) ,

a Fourier analysis is requested on a fundamental frequency of 500Hz at Node 6. PSpice's harmonic decomposition calculates the magnitude and phase of the fundamental and first eight harmonics (Fourier or spectral analysis) and automatically provides data for .PRINT, .PLOT, and Probe.

See Fourier analysis example in Chapter 1, EOC problem.

I Current Source, I(Name) (Node1) (Node2) (DC or AC) (Value),
I1 1 0 DC 1A

.IC

Initial Bias Point Calculations, presets the initial bias points in .TRAN and small-signal analysis. For example, .IC V(6)=2. The assigned node voltage is disengaged after bias point calculations are completed, after which the specified analysis is run. If both .IC and .NODESET are included in control statements, .IC overrides all .NODESET statements.

J Junction Field Effect Transistor, JFET,
J(Name) (Drain) (Gate) (Source) (Model) [See Appendix B]

K Coupling Coefficient. [See KTRANFRM and Using Transformers in PSpice in Chapter 8]

L Inductor, L(Name) (Node1) (Node2) (Value)
L1 1 0 10mH

.LIB

.LIB is a control statement directing the PSpice program to search the library file for device and subcircuit models used in the circuit. The .LIB EVAL.LIB entry is necessary to direct the program to search the Evaluation Library.
Ask instructor for a copy of the Evaluation Library file.

M Metal-Oxide Semiconductor Field Effect Transistor, MOSFET,
M(Name) (Drain) (Gate) (Source) (Substrate) (Model)
[See Appendix B]

.MC, .SENS, and .WCASE

Monte Carlo, Sensitivity, and Worst Case Analysis

With Monte Carlo analysis, component tolerances are randomly changed in multiple runs of .DC, .AC, or .TRAN analyses. The component values used in a particular run are printed along with the results of the analysis. In Monte Carlo all parameters under test are randomly varied and the entire circuit is tested.

In Sensitivity, .SENS, as with Monte Carlo, the first run is made with components at their predicted or "best-case" values. But, unlike Monte Carlo, .SENS analysis checks the circuit with only one component value changed. The output reveals the behavior of the circuit with this one component adjusted. Components tested include; resistors, diodes, and BJTs.

In any design, circuit component values and model parameters vary within tolerances. These variations will change design bias points and overall circuit operation. These changes may or may not be acceptable within specifications that dictate range values. The .SENS statement, a sensitivity analysis of the circuit, provides an output table of the DC sensitivity to variations in components to input voltages and currents. The output displays each component's contribution to variations in the selected <output value>. The component that has the greatest effect on the circuit's sensitivity can then be targeted for fine tuning to specs. It should be noted that large quantities of data are presented in this output with large numbers of components.

Worst Case, .WCASE, is a final sensitivity run with extremes of component tolerance. The user defines the extreme high or low setting.

.MODEL

.Model (Name)

A set of parameters that defines the particular component or device to be simulated and analyzed.

Models include:

Abbreviation	Device or Component
CAP	Capacitor
D	Diode
IND	Inductor
NPN	NPN Bipolar Transistor
NJF	N-Channel Junction Field Effect Transistor
NMOS	N-Channel MOSFET
PJF	P-Channel Junction Field Effect Transistor
PMOS	P-Channel MOSFET
PNP	PNP Bipolar Transistor
RES	Resistor

Device examples are:

.MODEL D1N4001 D(parameters...) defines the diode model that is called to be used in the circuit.

.MODEL Q2N2222 NPN(parameters....) calls for the 2N2222 BJT model in the Device Library. [See Appendix B]

.NODESET

Nodeset sets the initial bias point voltage level of the node specified. For example, .NODESET V(4)=4 provides a predetermined initial estimate of 4Volts to Node 4, for the analysis run. In .DC sweep, Nodeset disengages after the first step. Nodeset is also used in .TRAN and .AC analysis. In digital simulation, Nodeset can be used to set the initial condition of flip-flops. .IC statements override Nodeset.

.NOISE

.AC small signal analysis is necessary for .NOISE to be run. .Noise statements provide noise analysis of the particular part of a circuit that feeds the "output" of interest. For example, .NOISE V(6) Vin 2 0, specifies a noise analysis of the circuit that outputs at V(6). The input voltage, Vin, is the voltage source used in the small- signal analysis. The input voltage or current source used in the small signal analysis and circuit gains will be used in .Noise analysis. All noise contributors, resistors and devices, are taken into calculations. If a specific print interval is requested, .plot or .print outputs for each noise contributor at each analysis frequency are provided in volt/hertz or ampere/hertz units.

.OPTIONS

.OPTIONS Name
Used to set all options. Numerous are available.

Options Examples	Use
Acct	Include accounting information in output
List	Provides a list of components and devices
Noecho	Circuit netlists are usually included in the output file. This option excludes netlists.
Nopage	Groups the output file into a continuous scroll-type presentation, instead of separate pages for each section of the report.
GMIN	Sets the minimum conductance of any branch.
Reltol	Sets relative accuracy of voltage and current.

.PLOT

This command outputs DC/AC/NOISE and transient analysis to plots.
.PLOT graphs values in the Output file. This command will graph any type sweep and the swept values are presented on the Y-axis with the input sweep values (i.e., Vin) on the X-axis. It is a good idea to limit the number of plot

points, thereby limiting the size of graphs. If no range is entered, default is the full range of swept values. The graphed plots use keyboard symbols.

.PRINT

Outputs DC/AC/NOISE and transient analysis in tables.

.PROBE

Writes DC/AC/NOISE and transient analysis outputs to a data file.
A control statement directing PSpice to save all data from the analysis for use by Probe. If no values are specified after .Probe, all node voltages and device currents will be save in a data file. The data filename will have a .DAT extension. Extensive data can be generated with the .PROBE command.

Q Bipolar Junction Transistor;
Q(Name) (Collector) (Base) (Emitter) (Model) [See Appendix B]

R Resistor, R(Name) (Node1) (Node2) (Value)
R1 1 0 100K

.SENS (See Monte Carlo)

.SUBCKT Subcircuit, .Subckt(Name) (Node listings...) [See Appendix B]

.TEMP

Sets the temperature for the analysis run. For example, .TEMP 30 sets the temperature at 30 degrees C. Default setting is 27 degrees C. Temperature can be set at any test value, or swept in a DC sweep statement. For example, voltage and current sources and individual component values can be swept while stepping temperature levels. Probe is used to analyze the changes effected by the sweep.

.TRAN Transient_Analysis

Circuit response to an input stimulus (input signal) is measured over time with the transient analysis statement, .TRAN. The amplitude, frequency, phase, and other parameters of the input signal are set with a voltage input "generator" statement such as Vin 1 0 SIN(0 .1V 5KHz). This input amplitude is set to .1Vpeak, and frequency is 5KHz. The transient analysis statement specifies the analyses of the circuit with this stimulus during a specific period of time. A typical statement follows on the next page.

.TRAN .004ms .4ms 0 .002ms

The first entry in the transient analysis line is .004ms. It is the print_step time interval for .print and .plot results from transient analysis. This time is also stored for the Probe Graphical Waveform Analyzer program. The print_step time is calculated as final_time divided by 100 (.4ms/100= .004ms) for this analysis. Print_step and each of the other settings are adjusted for specific analysis at the discretion of the user.

Final_time is the next entry and is .4ms. It has been set to analyze two hertz of a 5KHz input. Final_time is calculated as 2Hz(t = 1/f) and is the actual time for the transient analysis to run.

The third entry is results_delay and is set to 0 time. It suppresses the transient analysis start for zero time (no delay).

Step_ceiling is set to .002ms and is the last entry in the .TRAN statement. The step_ceiling default setting is calculated as final_time/50, .4ms/50= .008ms, and is the maximum step_ceiling time for transient analysis. In this analysis final_time/200 is used for better Probe graphical displays. (.4ms/200=.002ms)

To repeat, the .TRAN statement simulates the operation of the circuit in time, specifically during 2 hertz at 5KHz of the input stimulus (input signal). If nonlinear operation has been introduced into the circuit by, for examples, setting the bias points, by overdriving the inputs, or by exceeding the frequency limitations of the circuit, the outputs displayed by Probe will reveal the nonlinearity.

Transient Analysis Examples:

(1) .TRAN .33m 33.2m 0 .16m, programs a transient analysis to be performed on the circuit. The time period of 1 hertz at 60Hz is 16.6ms. Final_time is set to 33.3ms for an analysis period of 2 hertz...the analysis run time. Print_step time interval is .33ms for .print and .plot results from transient analysis, and stored for Probe. The third entry is results_delay and is set to 0 time. This suppresses the transient analysis start for zero time (no delay). A step_ceiling is entered at .1ms. It sets the time of the transient analysis interval time step. Step_ceiling default is final time/50 or .66ms. The .16ms step_ceiling will provide more evenly drawn Probe plots.

(2) Transient_analysis will provide the analysis of 2 hertz at 1KHz in this example. The input is defined as,

Vin 1 0 SIN(0 10V 1KHz),

and the analysis time is defined in the .TRAN control statement,

.TRAN .02m 2m 0 .01m

Vin defines the type of input waveform (sine wave), the input voltage level (Vpeak=10V), and frequency (1KHz) for transient analysis, and other circuit measurements.

The .TRAN statement defines the transient time period in which to analyze the operation of the circuit; final time= 2ms. (2 cycles at 1KHz) It also sets print_step with the .02ms entry. Typically, the print step is set to 1/100th of the final time. Step_ceiling is limited by the last entry in this statement, .01ms. The default value would have been final time/50 = .04ms.

V Voltage Source, V(Name) (Node1) (Node2) (DC value) or (AC value)

Voltage Source Examples:

Vin 1 0 DC 12V, indicates a 12Volt DC voltage source connected at Nodes 1 and 0. If DC is omitted, PSpice default is DC.

Vin 5 0 AC 0.001V SIN(0 0.001V 5KHz) sets the AC input stimulus specifications and levels.

AC 0.001V identifies the small-signal analysis specification. Later in the circuit file, the .AC control statement places frequency sweep limits on the input as well as defines the data sample number.

SIN(0 0.001V 5KHz) sets the transient analysis input specs. The .TRAN statement defines the required analysis of the 5KHz signal. The entries in parenthesis are offset_voltage= 0V, input signal amplitude=.001V, and signal frequency= 5KHz. Three other entries are omitted and default setting are accepted. They are time_delay= 0 seconds, damping_factor= 0, and input signal phase adjustment= 0 degrees. Each of these parameters can be entered or changed with the Circuit Editor or with the Stimulus Editor (StmEd).

.WCASE (See Monte Carlo)

.WIDTH

.WIDTH OUT=80 sets the width of the output file to 80 columns. Two column settings are available, 80 and 132. If left unset, default is 80.

X Indicates a call for a particular subcircuit,
X(Name) (Node listings...) (Subcircuit name)

For example, XAMP 3 4 1 2 8 UA741, is the call for the 741 operational amplifier subcircuit. The node connection numbers correspond to circuit requirements. [See Appendix B]

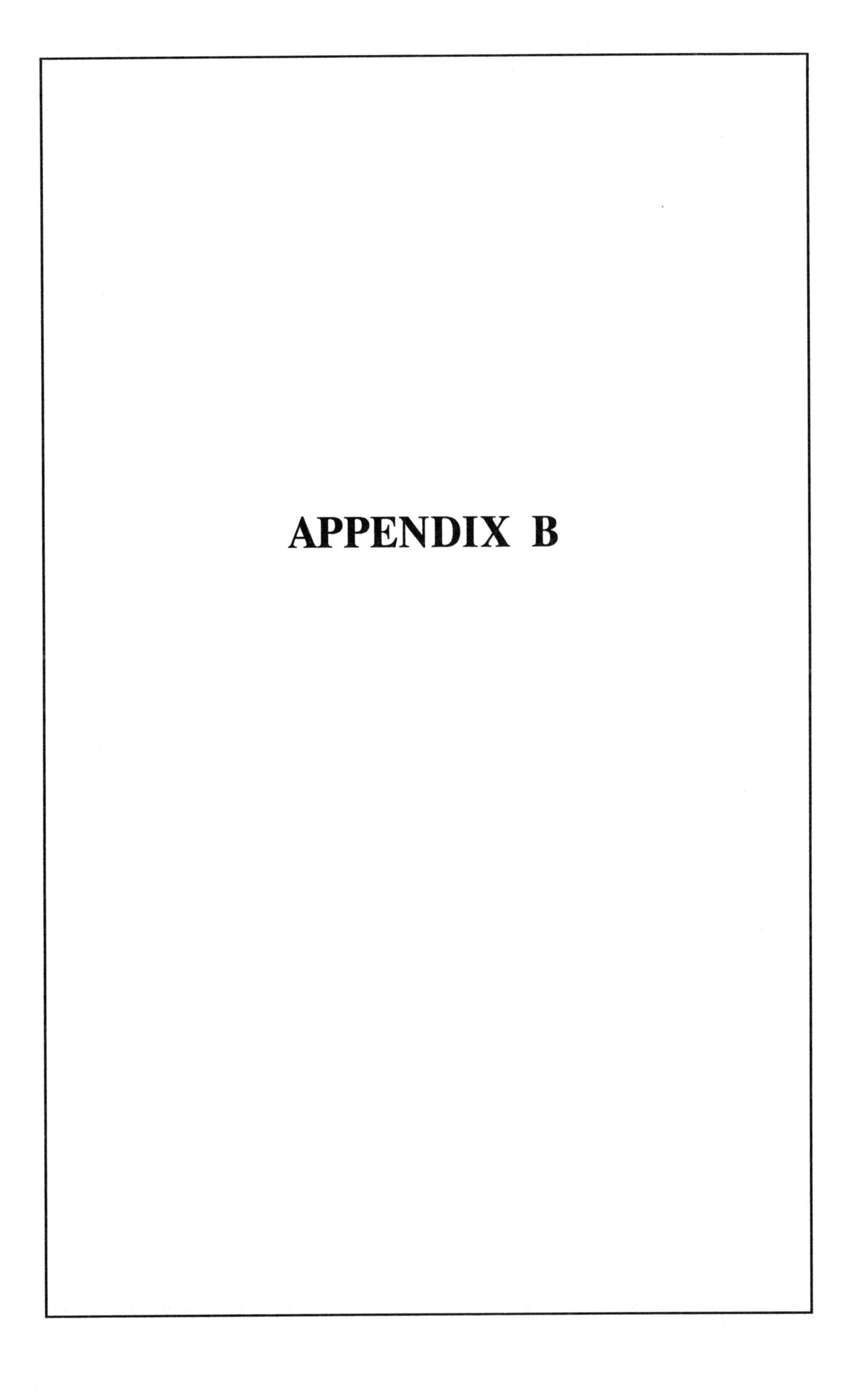

APPENDIX B

APPENDIX B

DEVICE MODEL PARAMETERS AND SUBCIRCUIT STATEMENTS

DEVICE MODELS

The following is a listing of model parameters and definitions. Default values are accepted when parameters are not listed. Print and review the EVAL.LIB file for additional model and subcircuit data.

BJT MODEL PARAMETERS

The following model and parameter list includes only a few of the more than 50 parameters used in BJT model definitions.

```
.Model  Q2N2222A  NPN(Is=14.34f Xti=3 Eq=1.11 Vaf=74.03
+                 Bf=255.9  Ne=1.307  Ise=14.34f  Ikf=.2847
+                 Xtb=1.5 Br=6.092  Nc=2 Isc=0  Ikr=0  Rc=1
+                 Cjc=7.306p  Mjc=.3416  Vjc=.75  Fc=.5
+                 Cje=22.01p  Mje=.377  Vje=.75  Tr=46.91n
+                 Tf=411.1p  Itf=.6  Vtf=1.7  Xtf=3  Rb=10)
*                 National        pid=19        case=TO18
*                 88-09-07        bam           creation
```

BJT Parameter	Definition
Is	Transport saturation current
Xti	IS temperature effect exponent
Eg	Bandgap voltage (barrier height)
Vaf	Forward Early voltage
Bf	Ideal maximum forward beta
Ne	Base-emitter leakage emission coefficient
Ise	Base-emitter leakage saturation current
Ikf	Forward-beta high-current rolloff "knee" current
Xtb	Forward and reverse beta temperature coefficient
Br	Ideal maximum reverse beta
Nc	Base-collector leakage emission coefficient
Isc	Base-collector leakage saturation current
Ikr	Corner for reverse beta high current rolloff
Rc	Collector ohmic resistance
	Cont.

BJT Parameter	Definition
Cjc	Base-collector zero bias p-n capacitance
Mjc	Base-collector p-n grading factor
Vjc	Base-collector built in potential
Fc	Forward-bias depletion capacitor coefficient
Cje	Base-emitter zero-bias p-n capacitance
Mje	Base-emitter p-n grading factor
Vje	Base-emitter built-in potential
Tr	Ideal reverse transit time
Tf	Ideal forward transit time
Itf	Transit time dependency on collector current, IC
Vtf	Transit time dependency on VBC
Xtf	Transit time bias dependency coefficient
Rb	Zero-bias (maximum) base resistance

DIODE MODEL PARAMETERS

POWER DIODE

```
.Model D1N4001 D(Is=10.0E-15 Rs=.1 Ikf=0 Cjo=1p N=1
+                Eg=1.11 Xti=3 Vj=.75 Fc=.5 Nr=2 Bv=100
+                Ibv=100.0E-6  Tt=5n Isr=100.0E-12)
```

NOTE: The 1N4001 model is not a part of the EVAL.LIB file. If needed, a model must be written in the circuit netlist. Be sure to include the "+" to begin the additional lines.

ZENER DIODE

```
.Model D1N750 D(Is=880.5E-18 Rs=.25 Ikf=0 N=1 Xti=3 Eg=1.11
+               Cjo=175p M=.5516 Vj=.75 Fc=.5 Isr=1.859n  Nr=2
+               Bv=4.7  Ibv=20.245m  Nbv=1.6989  Ibvl=1.9556m
+               Nbvl=14.976  Tbvl=-21.277u)
```

Diode Parameter	Definition
Is	Saturation current
Rs	Parasitic resistance
Ikf	High-injection "knee" current
Cjo	Zero-bias p-n capacitance
N	Emission coefficient
Eg	Bandgap voltage (barrier height)
	Cont.

Diode Parameter	Definition
Xti	IS temperature exponent
Vj	P-N potential
Fc	Forward-bias depletion capacitance coefficient
Nr	Emission coefficient for ISR
Bv	Reverse breakdown "knee" voltage
Ibv	Reverse breakdown "knee" current
Tt	Transit time
Isr	Recombination current parameter
M	P-N grading coefficient
Nbv	Reverse breakdown ideality factor
Ibvl	Low-level reverse breakdown "knee" current
Nbvl	Low-level reverse breakdown ideality factor
Tbvl	BV temperature coefficient (linear)

JFET MODEL PARAMETERS

```
.Model J2N3819 NJF(Beta=1.304m Betatce=-.5 Lambda=2.25m
+          Vto=-3, Vtotc=-2.5m  Is=33.57f  Isr=322.4f
+          N=1 Nr=2 Xti=3 Alpha=311.7  Vk=243.6
+          Cgd=1.6p  M=.3622 Pb=1  Fc=.5 Cgs=2.414p
+          Kf=9.882E-18  Af=1)
```

JFET Parameter	Definition
Beta	Transconductance coefficient
Betatce	Beta exponential temperature coefficient
Lambda	Channel-length modulation
Vto	Threshold voltage
Vtotc	VTO temperature coefficient
Is	Gate p-n saturation current
Isr	Gate p-n recombination current parameter
N	Gate p-n emission coefficient
Nr	Emission coefficient for ISR
Xti	IS temperature coefficient
Alpha	Ionization coefficient
Vk	Ionization "knee" voltage
Cgd	Zero-bias gate-drain p-n capacitance
M	Gate p-n grading coefficient
Pb	Gate p-n potential
Fc	Forward-bias depletion capacitance coefficient
	Cont.

JFET Parameter	Definition
Cgs	Zero-bias gate-source p-n capacitance
Kf	Flicker noise coefficient
Af	Flicker noise exponent

MOSFET MODEL PARAMETERS

```
.Model IRF150 NMOS(Level=3 Gamma=0 Delta=0 Eta=0 Theta=0
+                  Kappa=0  Vmax=0  Xj=0 Tox=100n  Uo=600 Phi=.6
+                  Rs=1.624m  Kp=20.53  W=.3 L=2u  Vto=2.831
+                  Rd=1.031m  Rds=444.4K  Cbd=3.229n  Pb=.8
+                  Mj=.5 Fc=.5 Cgso=9.027n  Cgdo=1.679n
+                  Rg=13.89  Is=194E-18  N=1 Tt=288n)
```

MOSFET Parameter	Definition
Level	Model index
Gamma	Bulk threshold parameter
Delta	Width effect on threshold
Eta	Static feedback (Level=3)
Theta	Mobility modulation (Level=3)
Kappa	Saturation field factor (Level=3)
Vmax	Maximum drift velocity
Xj	Metallurgical junction depth
Tox	Oxide thickness
Uo	Surface mobility
Phi	Surface potential
Rs	Source ohmic resistance
Kp	Transconductance coefficient
W	Channel width
L	Channel length
Vto	Zero-bias threshold voltage
Rd	Drain ohmic resistance
Rds	Drain-source shunt resistance
Cbd	Zero-bias bulk-drain p-n capacitance
Pb	Bulk p-n bottom potential
Mj	Bulk p-n bottom grading coefficient
Fc	Bulk p-n forward-bias capacitance coefficient
Cgso	Gate-source overlap capacitance per unit channel width
Cgdo	Gate-drain overlap capacitance per unit channel width
Rg	Gate ohmic resistance
Is	Bulk p-n saturation current
N	Bulk p-n emission coefficient
Tt	Bulk p-n transit time

UA741 OPERATIONAL AMPLIFIER

UA741 SUBCIRCUIT

```
* connections:     non-inverting  input
*                  | inverting  input
*                  | |  positive  power  supply
*                  | | |  negative  power  supply
*                  | | | |  output
*                  | | | | |
.subckt  UA741  1 2 3 4 5
c1   11  12  8.661E-12
c2    6   7  30.00E-12
dc   5  53  dx
de   54  5   dx
dlp   90  91  dx
dln   92  90  dx
dp   4  3  dx
egnd  99  0  poly  (2)  (3,0)  (4,0)  0  .5  .5
fb   7  99  poly(5)   vb   vc   ve   vlp   vln  0  10.61E6   -10E6  10E6
+           10E6  -10E6
ga   6  0  11  12  188.5E-6
gcm   0  6  10  99  5.561E-9
iee   10  4  dc  15.16E-6
hlim  90  0  vlim   1k
q1   11  2  13  qx
q2   12  1  14  qx
r2   6  9  100.0E3
rc1   3  11  5.305E3
rc2   3  12  5.305E3
re1  13  10  1.836E3
re2  14  10  1.836E3
ree  10  99  13.19E6
ro1  8  5  50
ro2  7  99  100
rp  3  4  18.16E3
vb  9  0  dc  0
vc  3  53  dc  1
ve  54  4  dc  1
vlim   7  8  dc  0
vlp  91  0  dc  40
bln   0  92  dc  40
.model   dx  D(Is=800.0E-18   Rs=1)
.model   qx  NPN(Is=800.0E-18   Bf=93.75)
.ends
```

Note the UA741 subcircuit listing begins with .subckt and concludes with .ends.

OP AMP CIRCUIT CONNECTIONS EXAMPLE

XAMP 3 4 1 2 8 UA741 defines the UA741 op amp subcircuit call. Parameter data, from .subckt to .ends statements, are drawn from the library file for this model and listed on the previous page. The control statement is very similar to a macro line. The numbers define the UA741 model connections by position vs use:

Position	Node#	Use	
First position	3	Noninverting input connection	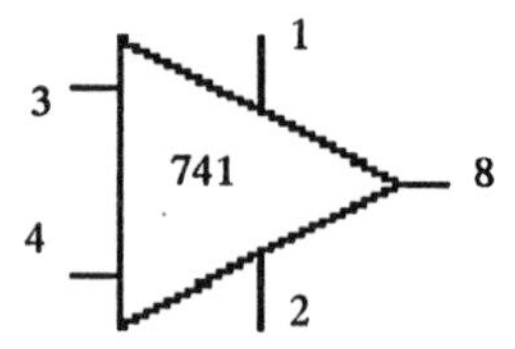
Second position	4	Inverting input	
Third position	1	Positive voltage source	
Fourth position	2	Negative voltage source	
Fifth position	8	Output	

Note the numbers used in the circuit are not necessarily the position numbers. A designated node connection on the op amp must correspond with the position in the library model. Any node number may be used if that number is placed in the proper position for the model.